CAMBRIDGE O LEVEL
BIOLOGY
With Stafford

Paper Reference: 5090

DEDICATION

This book is dedicated to **Mrs. T. Ambrose**, my first and most beloved
Biology teacher in Stewart School, Bhubaneswar.
Thank you Ma'am for your dedication and sincerity.

Typeset & layouts by : Mohamed Sobir
Cover designed by : Mohamed Sobir
Printed in Maldives by : Copier Repair
Printed Internationally by : Amazon.com
Published by : Author publisher
Date : June 2015

ISBN: 978-81-910705-8-3

Contact details of the author:

Stafford Valentine Redden
Near Hindhustan Oil Mill,
Ramchandrapur, Jatni P.O,
Orissa, India, 752050
staffordv@yahoo.com

CONTENTS

Chapter One
Cell Structure and Organisation

Cambridge 5090 syllabus specification 1(a) examine under the microscope an animal cell (e.g. from fresh liver) and a plant cell (e.g. from Elodea, a moss, onion epidermis, or any suitable, locally available material), using an appropriate temporary staining technique, such as iodine or methylene blue;
Cambridge 5090 syllabus specification 1(b) draw diagrams to represent observations of the plant and animal cells examined above;
Cambridge 5090 syllabus specification 1(c) identify, from fresh preparations or on diagrams or photomicrographs, the cell membrane, nucleus and cytoplasm in an animal cell;
Cambridge 5090 syllabus specification 1(d) identify, from diagrams or photomicrographs, the cellulose cell wall, cell membrane, sap vacuole, cytoplasm, nucleus and chloroplasts in a plant cell;

A cell is the basic structural and functional unit of life. All living things are made up of cells.

Most cells are very small and have to be seen by using a microscope. However, even with the microscope the parts of the cell are not clearly visible. To make the structures within a cell more visible, the cells must be **coloured or stained** with a suitable dye.

Two commonly used dyes are **Iodine solution** and **methylene blue**.

Iodine is commonly used to stain plant cells, like onion epidermal cells. Iodine makes parts of the cell which were translucent, more visible. This is because iodine stains starch present in the onion cell.

Methylene blue - stains cells to make their nuclei more visible.

The slide shown in figure 1.1 indicates a few structures that are visible in plant cells with a light microscope.

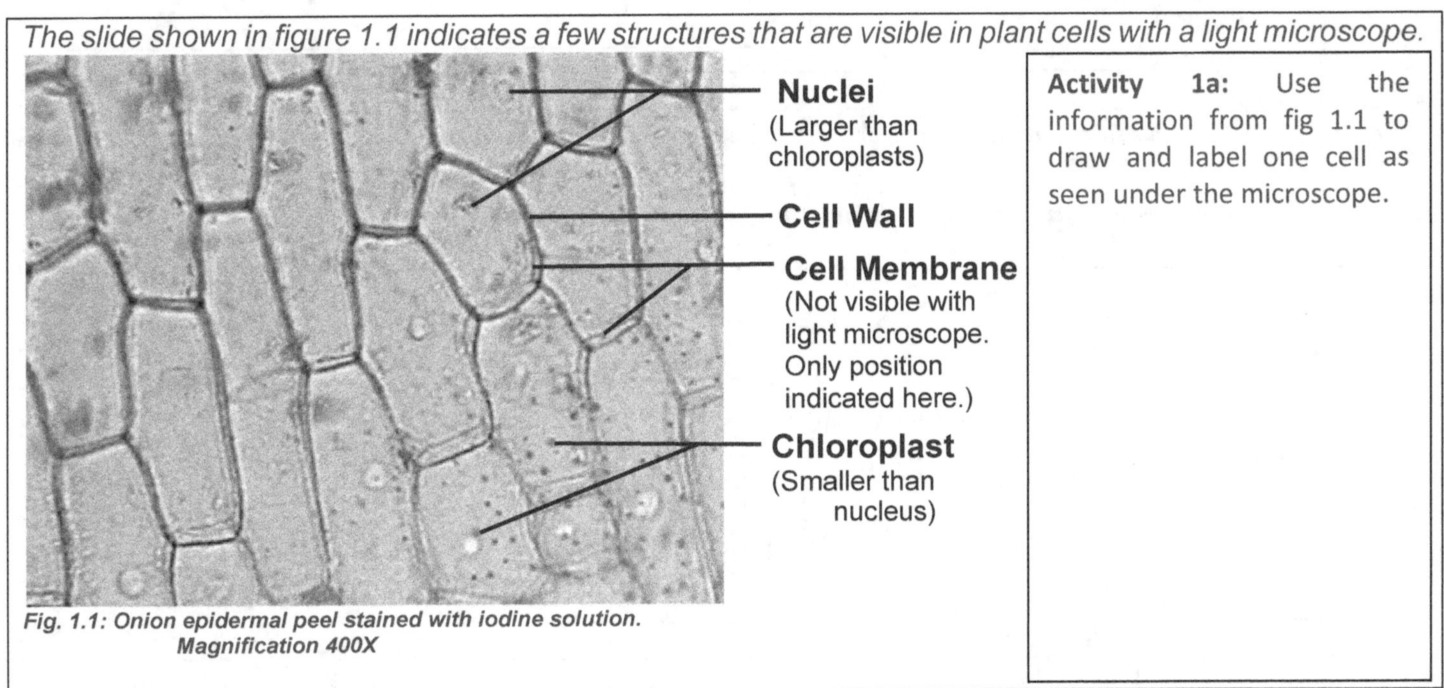

Fig. 1.1: Onion epidermal peel stained with iodine solution.
Magnification 400X

Activity 1a: Use the information from fig 1.1 to draw and label one cell as seen under the microscope.

"

The slide shown in figure 1.2 indicates a few structures that are visible in moss cells with a light microscope.

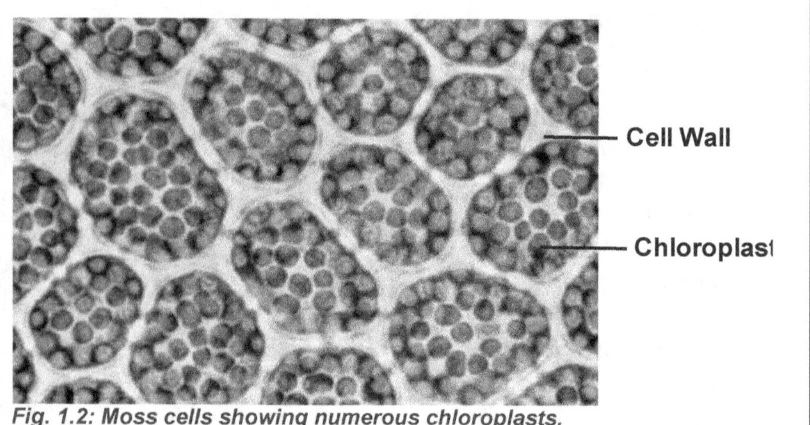

Fig. 1.2: Moss cells showing numerous chloroplasts.
Magnification 600X

Activity 1b: Calculate the actual size of a single cell shown in figure 1.2. Use the equation

$$Magnification = \frac{Size\ of\ image}{Size\ of\ object}$$

The slide shown in figure 1.3 indicates a few structures that are visible in liver cells with a light microscope.

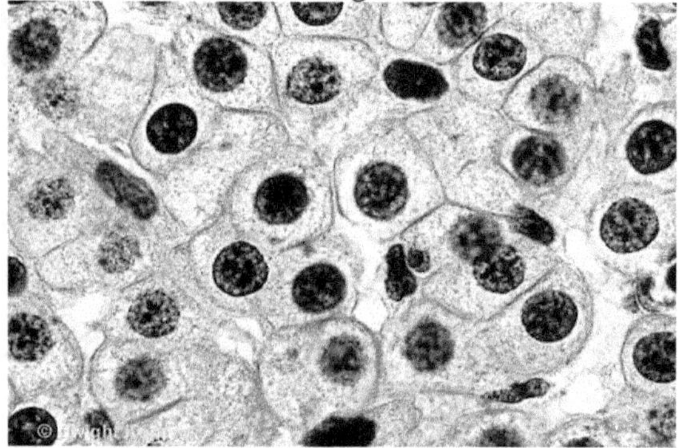

Fig. 1.3: Liver cell showing nucleus.
Magnification 400X

Activity 1c: Use the information from fig 1.3 to draw one liver cell as seen under the microscope. Label the nucleus, cell membrane and cytoplasm on your diagram.

Cambridge 5090 syllabus specification 1(e) compare the visible differences in structure of the animal and the plant cells examined;

Visible differences between plant and animal cells

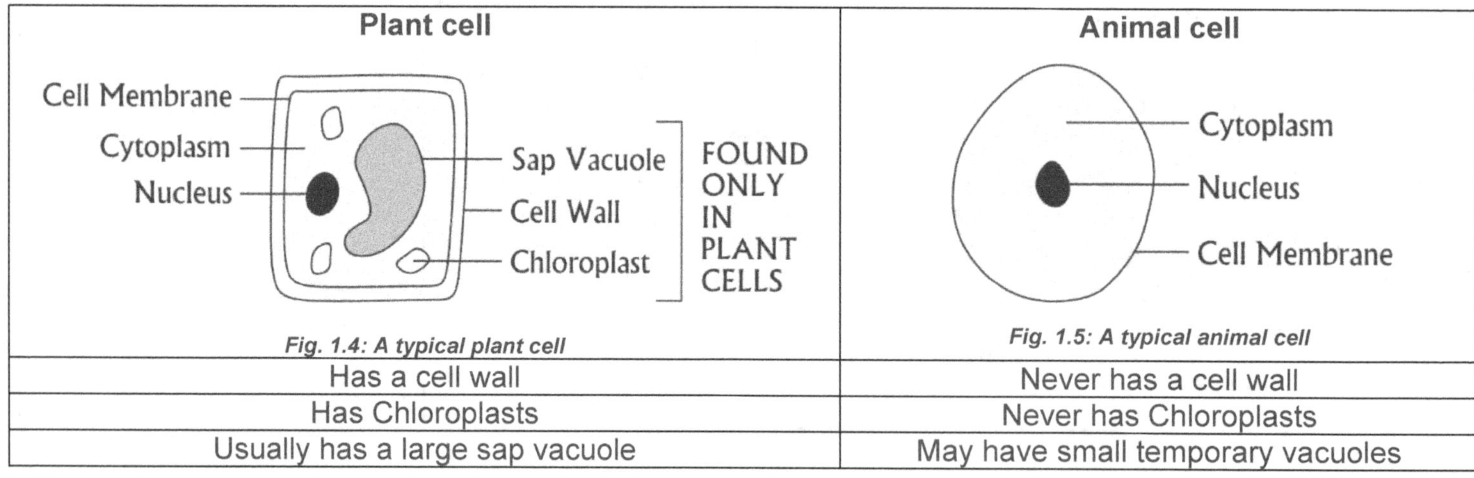

Fig. 1.4: A typical plant cell

Fig. 1.5: A typical animal cell

Plant cell	Animal cell
Has a cell wall	Never has a cell wall
Has Chloroplasts	Never has Chloroplasts
Usually has a large sap vacuole	May have small temporary vacuoles

Cambridge 5090 syllabus specification 1(f) state the function of the cell membrane in controlling the passage of substances into and out of the cell;
(g) state the function of the cell wall in maintaining turgor (turgidity) within the cell;

Cell membrane and cell wall

The cell membrane is the cell boundary and encloses the cytoplasm. The cell membrane is a selectively permeable membrane. This means that it will allow specific substances to pass through and will prevent other substances from entering or leaving the cell.

Water and gases can diffuse through the cell membrane freely. However, larger molecules and ions (solutes) are selectively absorbed or released from the cell.

The cell wall of plant cells is composed mainly of cellulose. The main function of the cell wall in the soft tissues of plants is to maintain turgor (turgidity) within the cell.

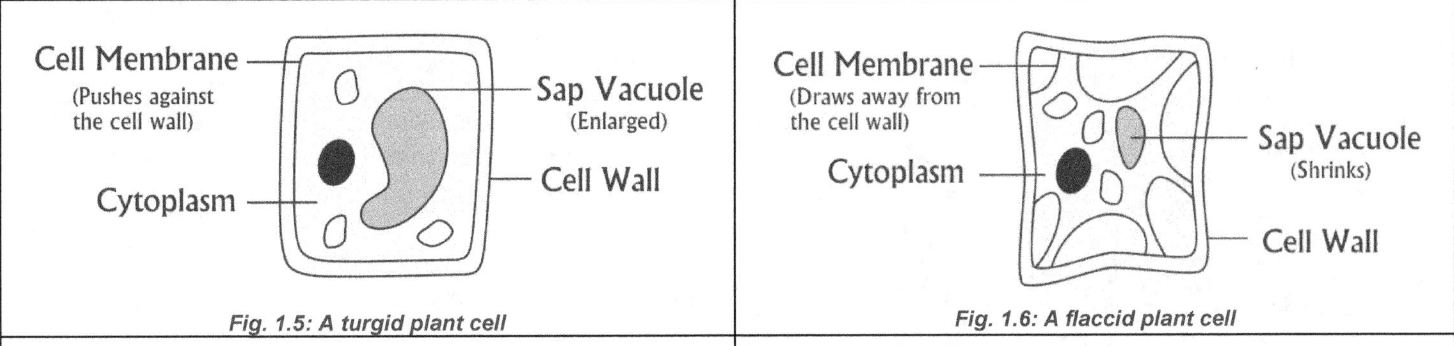

Fig. 1.5: A turgid plant cell

Fig. 1.6: A flaccid plant cell

A plant cell becomes turgid when the cell membrane pushes against the cell wall. This makes the cell become firm or stiff and provides mechanical support to soft tissues in plants. It is usually a result of endosmosis, where water enters the cell from the surroundings.	A plant cell becomes flaccid when the cell membrane draws away from the cell wall. The cell becomes soft and the cell is said to be plasmolysed. This is usually seen when plants lose excess water from the cells by exosmosis.

The **cell wall** prevents plant cells from bursting when water enters the cell by osmosis.
Animal cells will burst as the cell absorbs water, as there is no cell wall to provide support to the cell membrane.

In xylem and a few other cells in plants, the cell wall also contains a very hard, waterproof, woody substance called lignin to provide physical support to plants.

Cambridge 5090 syllabus specification 1(h) state, in simple terms, the relationship between cell function and cell structure for the following: • absorption – root hair cells; • conduction and support – xylem vessels; • transport of oxygen – red blood cells;
Cambridge 5090 syllabus specification 1(i) identify these cells from preserved material under the microscope, from diagrams and from photomicrographs;

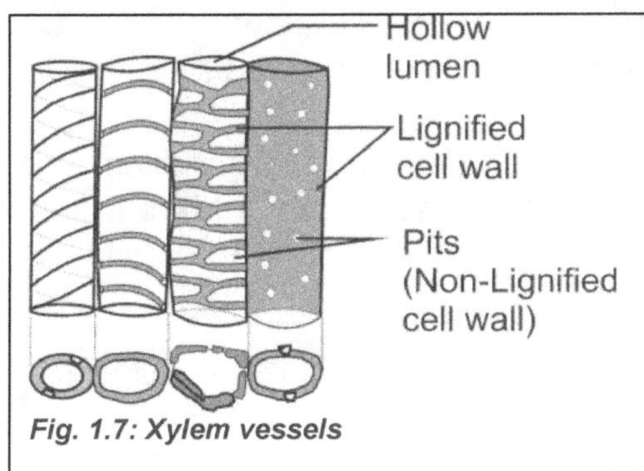

Fig. 1.7: Xylem vessels

Xylem vessels

- These are hollow tubes with no cytoplasm, no nucleus, no cell membrane, no cross walls. This allows sap to flow without any obstruction.
- They have highly lignified cell walls, with lateral pits. The lignin provides mechanical support. Lateral pits allow lateral movement of water and minerals from one vessel to another.
- They form a continuous system of tubes from roots to all parts of the plant. This allows them to transport water and ions from the roots to all parts of the plants.

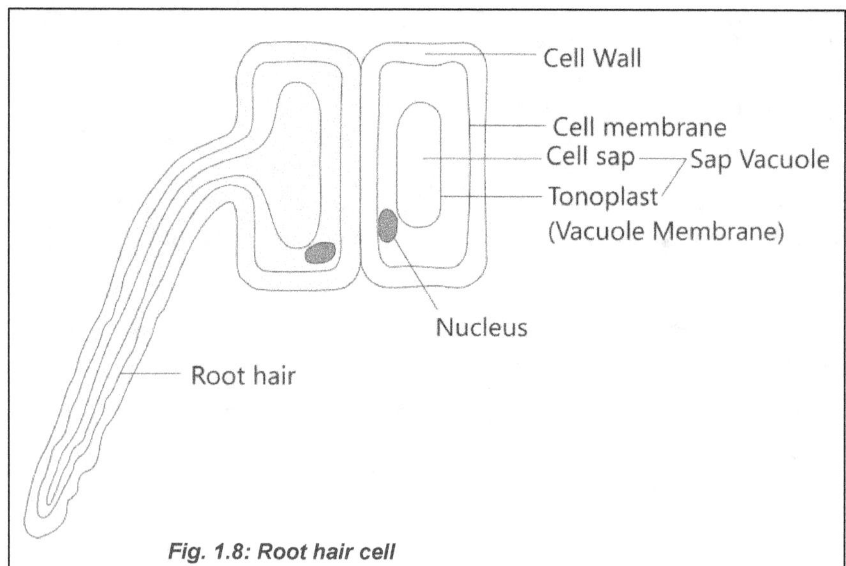

Fig. 1.8: Root hair cell

Root hair cell

- The root hair cell has an extension (root hair) which increases the surface area to volume ratio of the cell for rapid absorption of mineral ions, water and oxygen from the soil.

- The root hair is thin and has a low diffusion distance for rapid absorption of mineral ions, water and oxygen from the soil.

- The large sap vacuole helps to store ions and other solutes within the cell. It also stores excess water.

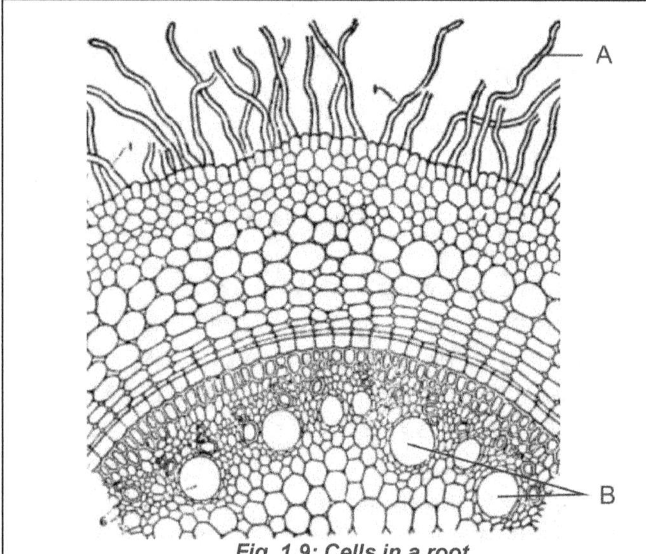

Fig. 1.9: Cells in a root

Activity 1d: Use the information from fig 1.9 to identify cells A and B. State one feature that helped to identify each tissue.

A ....

Reason:..

....

....

B ....

Reason:..

....

....

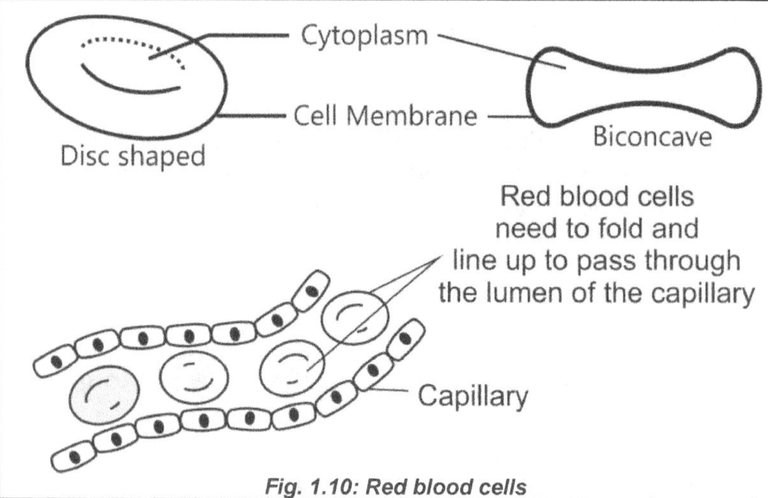

Fig. 1.10: Red blood cells

Red blood cell

- The red blood cell lacks a nucleus and other cell organelles. This provides more space to hold haemoglobin, which carries oxygen and carbon dioxide.

- The biconcave disc shape increases the surface area to volume ratio and decreases the diffusion distance to allow rapid diffusion of gases into and out of the red blood cells.

- The lack of a nucleus and organelles makes the cell pliable so that is can fold as it passes through the narrow lumen of the capillaries.

Cells, tissues, organs and organ systems

All living things are made up of cells. The cells may be organized into tissues, organs and organ systems. The levels of organization are shown in figure 1.11.

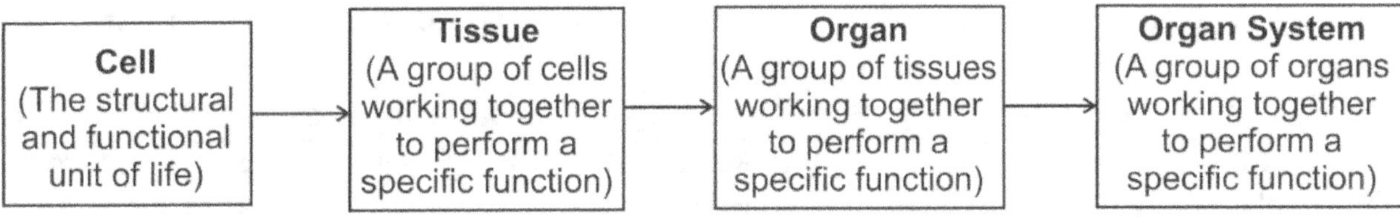

Fig. 1.11: Levels of organisation

The table below shows some of the cells commonly found in different tissues. It also shows the different tissues that make up organs. Some organs are also listed. It is **not necessary to memorise** all the names of cells, tissues or organs. However, it is necessary to be able to **distinguish** between a cell, tissue, organ and organ system.

Cells	Tissue	Organs	Organ system
Muscle cell, neurones, epithelial cells, etc.	Muscular tissue, Nervous tissue, connective tissue, etc.	Stomach, liver, pancreas, ileum, oesophagus, etc.	**Digestive System**
Epithelial cells, smooth muscle cells, etc.	Muscular tissue, connective tissue, blood, etc.	Kidneys, ureter, bladder, sphincter muscle, etc.	**Excretory system**
Neurones, glial cells, rods, cones, etc.	Nervous tissue, connective tissue, retinal tissue, etc.	Brain, spinal cord, eyes, ears, etc.	**Nervous system**
White Blood cells, red blood cells, muscle cells, etc.	Cardiac muscle tissue, blood, smooth muscle tissue, etc.	Heart, arteries, veins, lymph vessels, etc.	**Circulatory system**
Muscle cell, neurones, epithelial cells, etc.	Muscular tissue, Nervous tissue, connective tissue, blood, etc.	Ovaries, testes, penis, vagina, sperm duct, oviduct, uterus, etc.	**Reproductive system in humans**
Parenchyma, sieve tubes, sclerenchyma, vessels, collenchyma, etc.	Epidermal tissue, xylem, phloem, etc.	Anther, stigma, petal, sepal, receptacle, etc.	**Reproductive system in plants**

Magnification

Magnification is the ratio of size of the image to the actual size of the object.

$$Magnification = \frac{Size\ of\ image}{Size\ of\ object}$$

- If magnification is greater than 1 then it means that the image or drawing is larger than the object.
- If the magnification is less than 1 then it means that the image or drawing is smaller than the actual object.
- If magnification equals 1 then it means that the image is the same size as the object.

<table>
<tr><td>

Worked example

The photograph shows the leg of a cockroach.

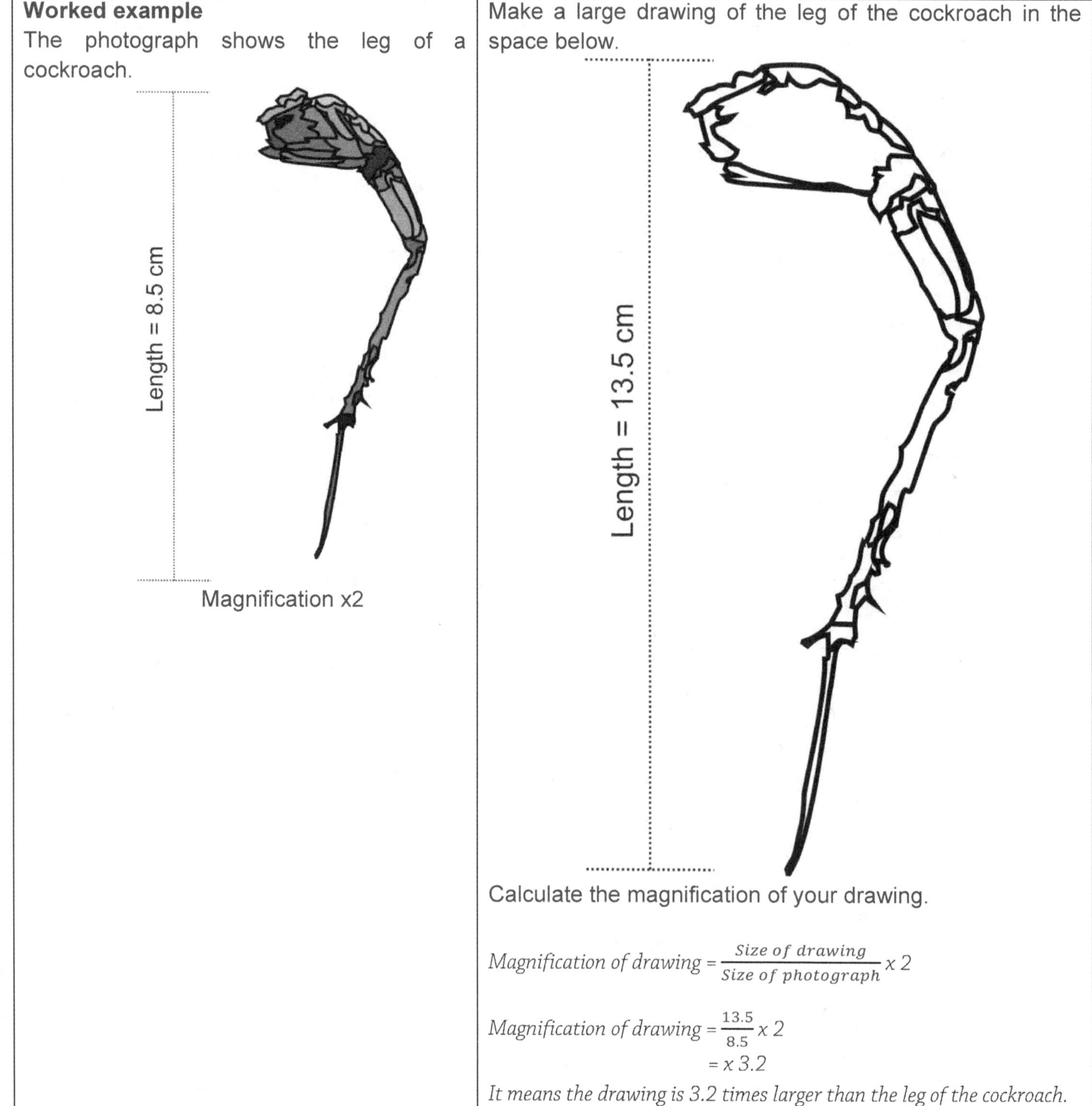

Length = 8.5 cm

Magnification x2

</td><td>

Make a large drawing of the leg of the cockroach in the space below.

Length = 13.5 cm

Calculate the magnification of your drawing.

$$Magnification\ of\ drawing = \frac{Size\ of\ drawing}{Size\ of\ photograph}\ x\ 2$$

$$Magnification\ of\ drawing = \frac{13.5}{8.5}\ x\ 2$$

$$= x\ 3.2$$

It means the drawing is 3.2 times larger than the leg of the cockroach.

</td></tr>
</table>

Cambridge 5090 syllabus specification 2(a) define diffusion as the movement of molecules from a region of their higher concentration to a region of their lower concentration, down a concentration gradient;

Diffusion

Diffusion is the net movement of particles **from a region of their higher concentration to a region of their lower concentration**, down a concentration gradient.

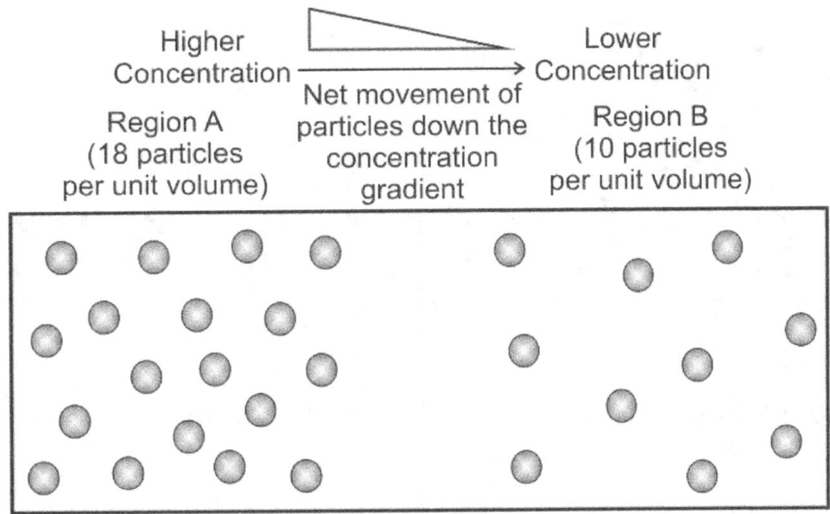

Fig. 2.1: Diffusion

Examples of diffusion
- Diffusion of oxygen from the air sacs (alveoli) into the blood, as shown in figure 2.2.
- Diffusion of carbon dioxide from the blood into the air sacs (alveoli), as shown in figure 2.2.
- Diffusion of water vapour from the air spaces in leaves to the external environment (transpiration).

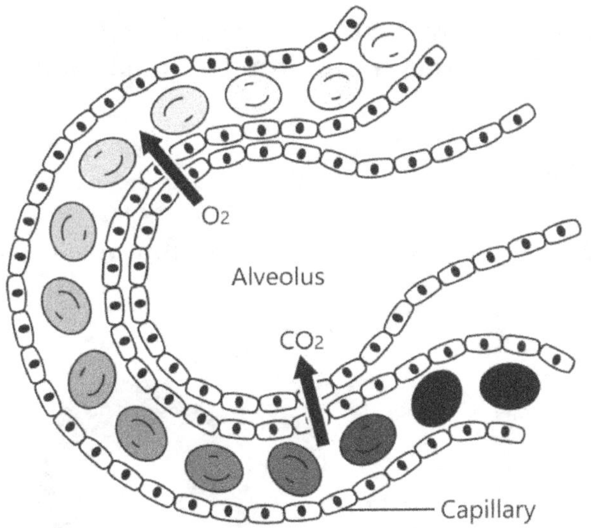

Fig. 2.2: Diffusion in lungs

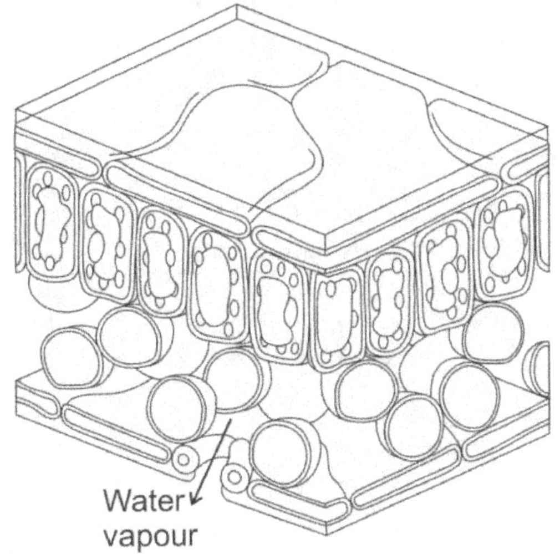

2.3: Diffusion in leaves

Cambridge 5090 syllabus specification 2(b) define osmosis as the passage of water molecules from a region of their higher concentration to a region of their lower concentration, through a partially permeable membrane;
Cambridge 5090 syllabus specification 2(c) describe the importance of a water potential gradient in the uptake of water by plants and the effects of osmosis on plant and animal tissues;

Osmosis

Osmosis is the movement of <u>**water molecules**</u> from a region of their higher concentration (higher water potential) to a region of their lower concentration (lower water potential), through a **partially permeable membrane**.

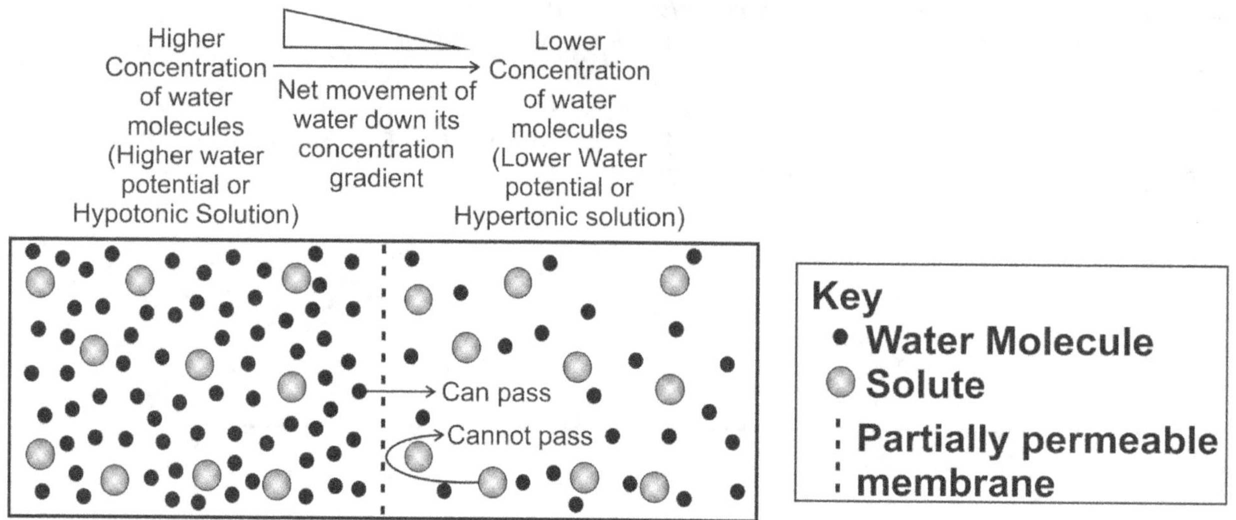

Fig. 2.4: Osmosis

A solution which has **lots of solute molecules** in a fixed volume of solvent is said to be highly concentrated and Hypertonic. It will have a **low water potential**.

A solution which has **few solute molecules** in a fixed volume of solvent is said to be at a lower concentration and Hypotonic. It will have a **high water potential**.

Two solutions which have the **same concentration of solutes** and the **same water potential** are said to be **isotonic**.

A partially permeable membrane will **allow water to pass through**, but will **restrict the movement of solute** molecules or ions. Examples of **partially permeable membranes** are **visking tubing**, **cell membrane** and the **inner membrane of an egg**.

The series of diagrams in figure 2.5 illustrates the effect of osmosis on red blood cells.

Red Blood Cell in a **hypertonic solution**	Red Blood Cell in an **isotonic solution**	Red Blood Cell in a **hypotonic solution or water**
Cell shrinks and shrivels (crenates)	No change in volume and shape of the cell	Cell swells and bursts (lysis)
H_2O moves out of the cell	No net movement of H_2O	H_2O moves in to the cell

Fig. 2.5: Effects of osmosis on red blood cells

The series of diagrams in figure 2.6 illustrates the effect of osmosis on plant cells.

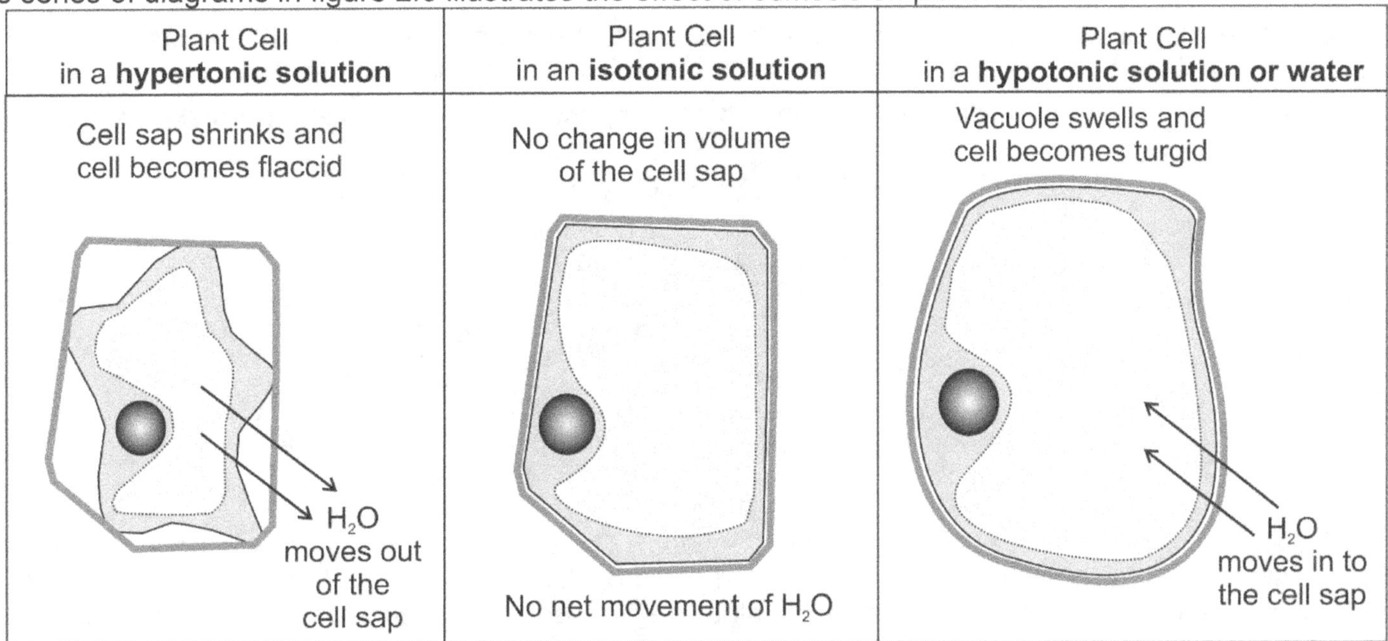

Fig. 2.6: Effects of osmosis on plant cells

- Plant cells do not burst due to the presence of a tough cell wall. The cell wall allows the cell to become firm and hard (turgid) when water enters the cell. This provides support to soft tissues in plants.
- Root hair cells absorb water by osmosis.

> *Cambridge 5090 syllabus specification 2(d) define active transport as the movement of ions into or out of a cell through the cell membrane, from a region of their lower concentration to a region of their higher concentration against a concentration gradient, using energy released during respiration.*
> *Cambridge 5090 syllabus specification 2(e) discuss the importance of active transport as an energy-consuming process by which substances are transported against a concentration gradient, as in ion uptake by root hairs and glucose uptake by cells in the villi.*

Active transport

Active transport is the movement of particles from **a region of their lower concentration to a region of their higher concentration** against a concentration gradient, using **energy** released **during respiration**. It occurs through specific proteins in the cell membrane.

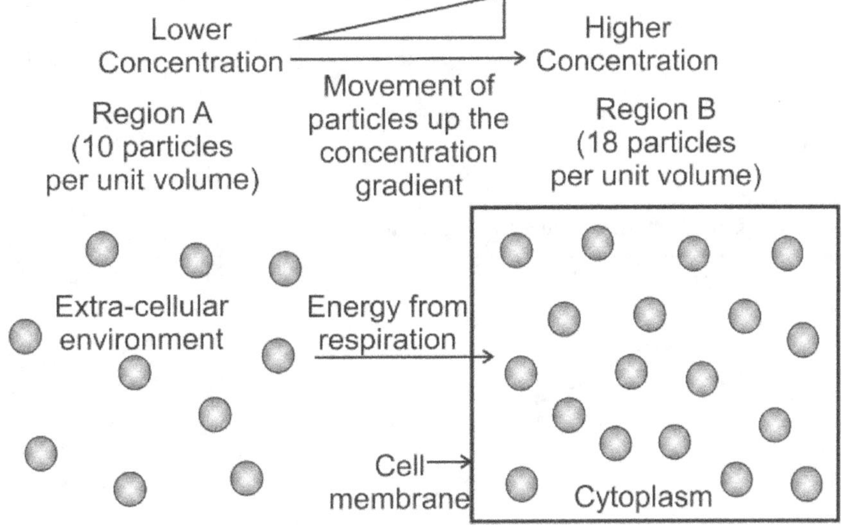

Fig. 2.7: Active transport

Cells which are involved in active transport usually have a **large surface area** and have a **high rate of respiration** to provide the **energy** needed **for active transport**.

Figure 2.8 shows a root hair cell which has a large surface area to volume ratio due to the extension (root hair). It also respires rapidly to provide the energy needed for active uptake of ions from the soil.

Note: Lack of oxygen in the soil due to waterlogging or compacted soil will reduce ion uptake as respiration slows down.

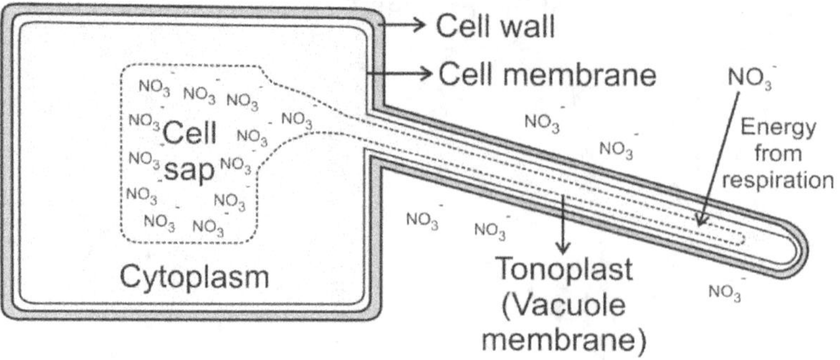

Fig. 2.8: Active transport of nitrate ions by root hair cells

Figure 2.9 shows an epithelial cell from the villi in the ileum. The cell has a large surface area to volume ratio due to the extensions (microvilli). It also respires rapidly to provide the energy needed for active uptake of ions from the soil. The glucose moves in to the cell against the concentration gradient. This ensures that glucose can be completely absorbed into the blood.

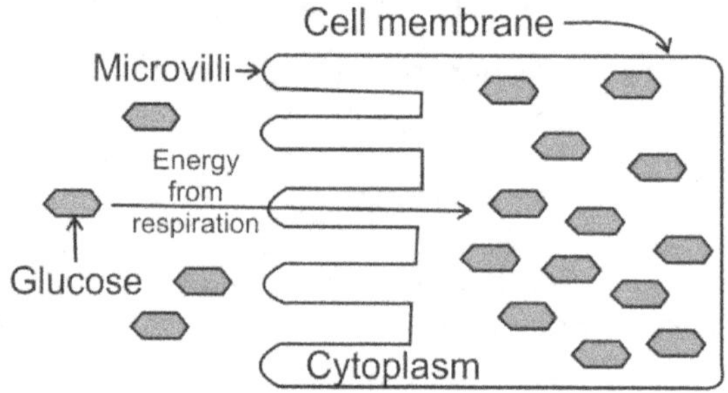

Fig. 2.9: Active transport of glucose in the ileum

Experiment to investigate osmosis in potato tissue

In an osmosis investigation, five pieces of raw potato of equal mass and a range of sucrose solutions of different concentrations was prepared. One piece of potato was placed in each sucrose solution. After two hours, the potato pieces were removed and blotted dry and the change in mass of each potato piece was calculated. The results are shown in the graph below.

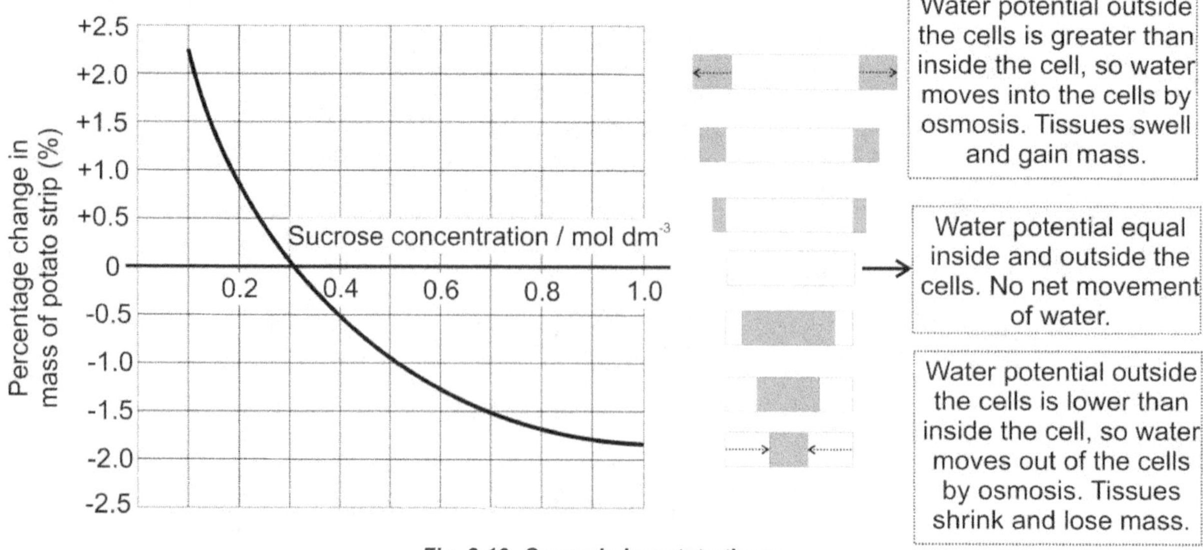

Fig. 2.10: Osmosis in potato tissue

Cambridge 5090 syllabus specification 3(a) define catalyst as a substance that speeds up a chemical reaction and is not changed by the reaction;
Cambridge 5090 syllabus specification 3(b) define enzymes as proteins that function as biological catalysts;

- Enzymes are proteins.
- Enzymes are **biological catalysts**. This means that they **speed up metabolic reactions** by lowering the activation energy.
- The enzymes **remains unchanged** at the end of the reaction.
- The substrate changes into the products.
- All enzymes react with a **specific substrate**. This is because the active site of every enzyme has a specific shape and can bind to a specific substrate only.

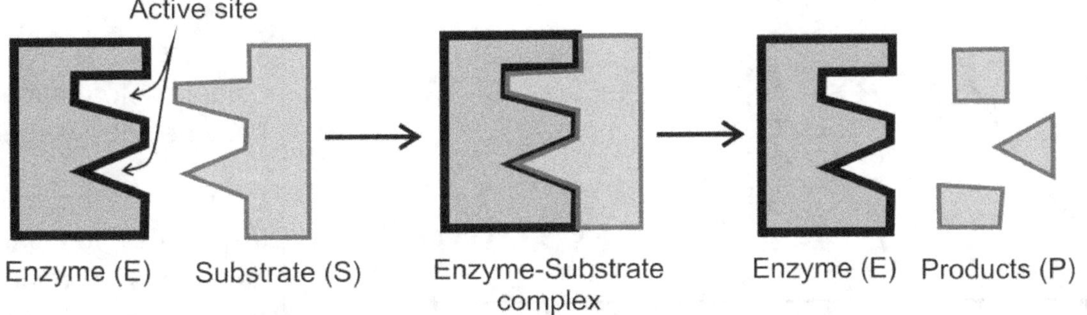

Fig. 3.1: Lock and key hypothesis of enzyme action

Cambridge 5090 syllabus specification 3(c) explain enzyme action in terms of the 'lock and key' hypothesis;

All enzymes are **soluble proteins** and have a **specific shape.** This specific shape is very essential for the functioning of enzymes. The part of the enzyme which reacts with the substrate is called the **active site**. The shape of the active site differs from one enzyme to another. This makes the enzyme (**lock**) react only with a specific substrate (**key**), which fits the active site.

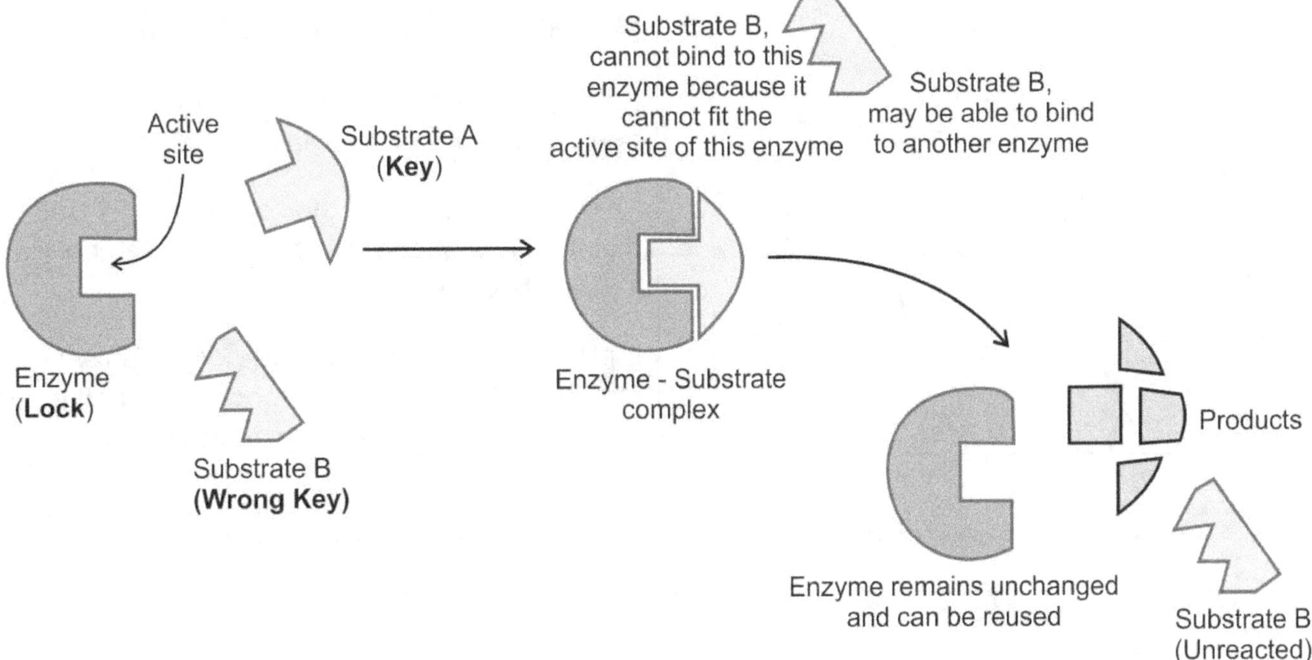

Fig. 3.2: Lock and key hypothesis of enzyme action (explains specificity)

Effect of temperature on the rate of enzyme activity

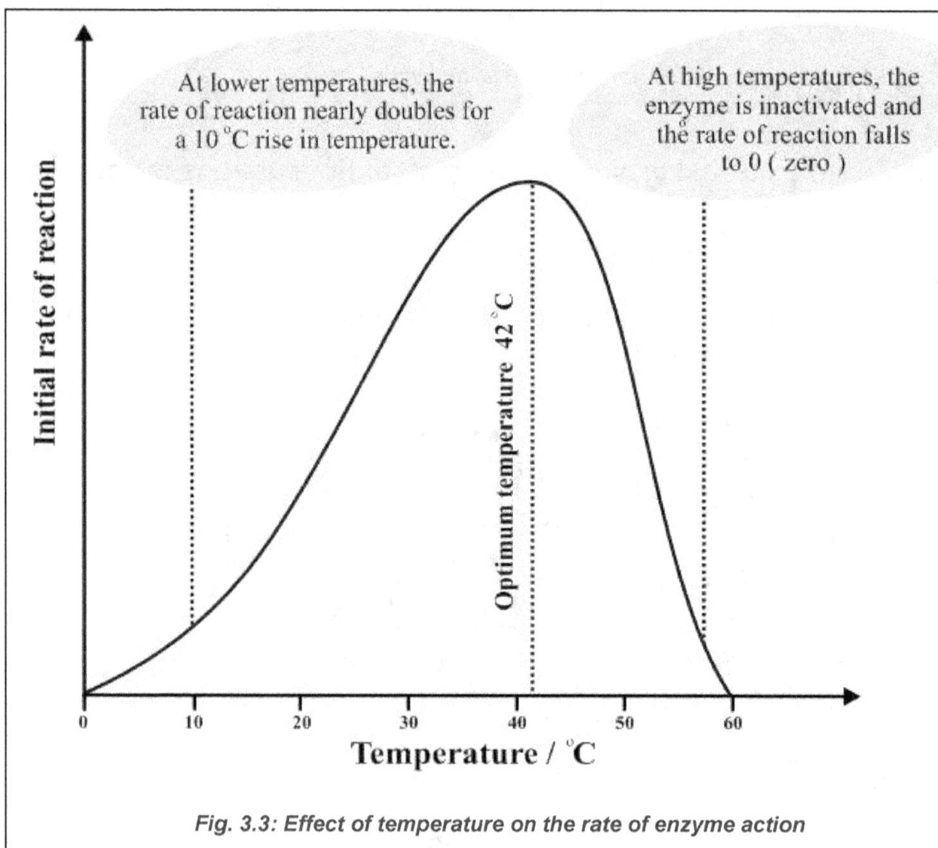

Fig. 3.3: Effect of temperature on the rate of enzyme action

As temperature increases up to the optimum, the rate of enzyme activity also increases.

This is because enzyme and substrate molecules gain more kinetic energy and the collisions between active sites of enzymes and substrate molecule become *more frequent*.

The *rate* of enzyme substrate complex formation increases so enzyme activity speeds up.

Beyond the optimum temperature, the rate of enzyme activity decreases because the high temperature causes the enzyme molecules to denature. The active site changes its shape and cannot bind with the substrate.

Effect of pH on enzyme activity

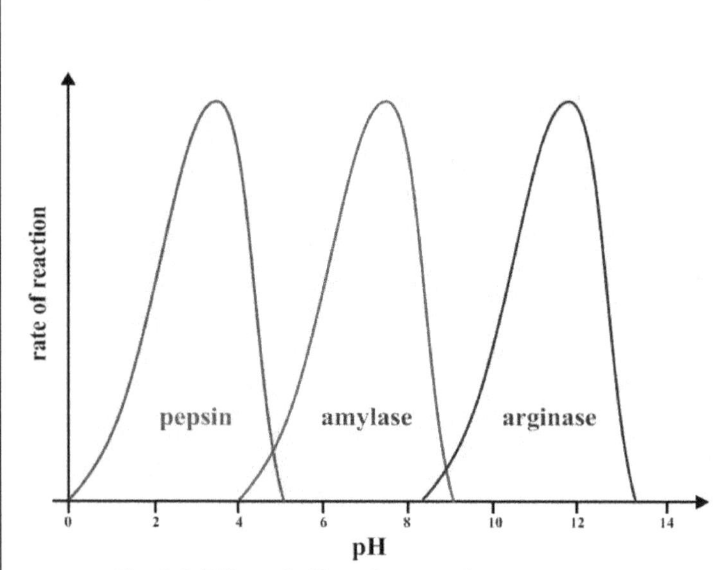

Fig. 3.4: Effect of pH on the rate of enzyme action

Enzymes are very sensitive to changes in pH.

Enzymes work best within a very narrow range of pH.

If the pH changes above or below the optimum then rate of reaction will decrease.

This is because at the optimum pH the enzyme maintains its shape and the substrate can bind to the active site of the enzyme.

If pH changes then the shape of the enzyme changes and the active site can no longer bind to the substrate. The rate of enzyme activity decreases.

Each enzyme will maintain its shape at a different pH. So, different enzymes have different optimum pH.

Investigating the effect of temperature on enzyme action

Independent variable: Temperature is varied.

Factors to be controlled (confounding variables)
- **pH**: the rate of enzyme action can be influenced by the pH of the solution. If the enzyme is subjected to an unsuitable pH, the enzyme denatures. So, the enzyme should be provided with the optimum pH to ensure reliable and valid results. A buffer solution of the optimum pH can be used to maintain an optimum pH of the enzyme solution. This makes the comparison fair and reliable.
- **Concentration of substrate**: the concentration of substrate should be kept constant for all trials. This makes the comparison fair and reliable.
- **Concentration of enzyme**: the concentration of enzyme should be kept constant for all trials. This makes the comparison fair and reliable.

Dependent variable: Rate of enzyme action

Step one: Place three test tubes with the enzyme, substrate and buffer solution in a water bath at 10⁰C for 10 minutes to allow the temperature to equilibrate, as shown in figure 3.5.

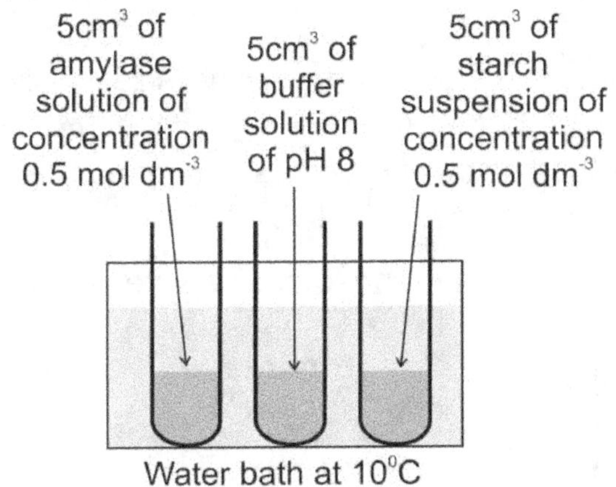

Fig. 3.5: Equilibration of temperature

Step two: Pour the enzyme solution and buffer solution into the test tube containing the starch suspension and place the mixture in the water bath at 10⁰C, as shown in figure 3.6.

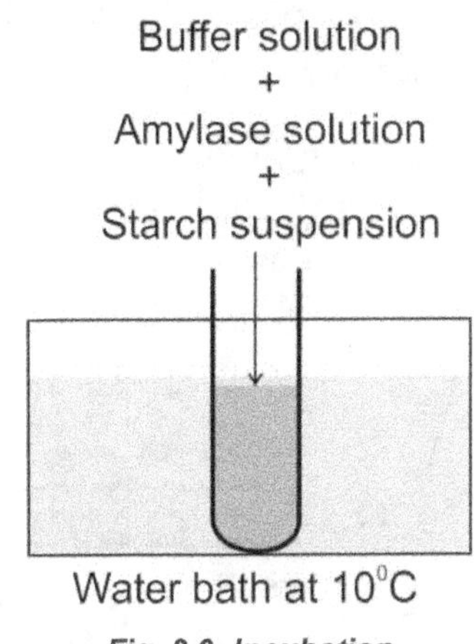

Fig. 3.6: Incubation

Step three: Stir the contents of the test tube and place a drop of the reaction mixture into a cavity tile containing 0.5 cm³ of iodine solution. Observe the colour change. Repeat the procedure at 2 minute intervals until the colour of the iodine does not change to blue-black. This is called the achromatic point, as shown in figure 3.7.

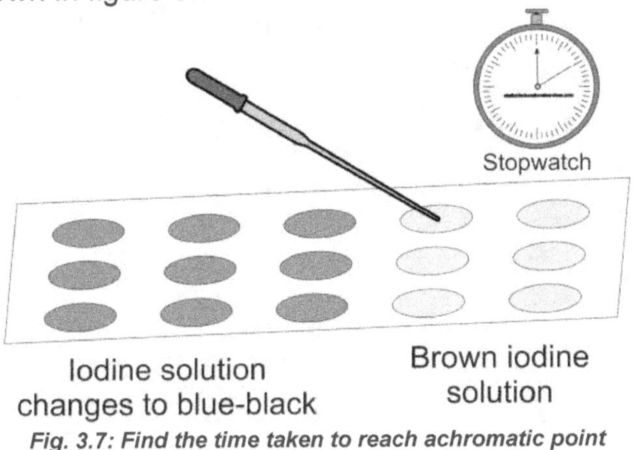

Fig. 3.7: Find the time taken to reach achromatic point

Step four: Determine the rate of reaction.

$$Rate = \frac{1}{Time\ taken\ to\ reach\ achromatic\ point}$$

Step five: Repeat the experiment at temperatures of 20⁰C, 30⁰C, 40⁰C, 50⁰C, 60⁰C, 70⁰C and 80⁰C.

Step six: Plot a graph to analyse the results.

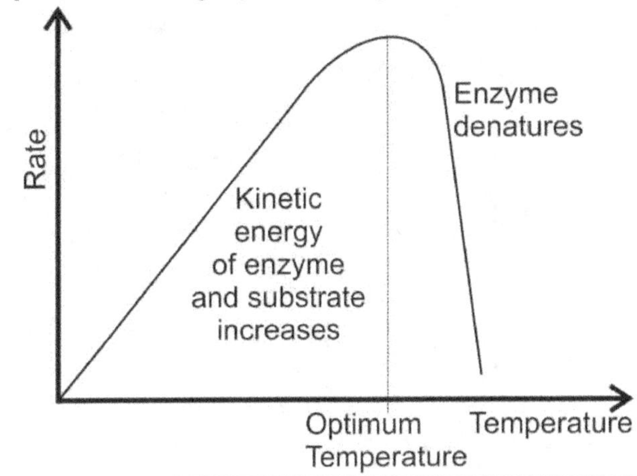

Investigating the effect of pH on the rate of enzyme action

Independent variable: pH is varied.

Factors to be controlled (confounding variables)

- **Temperature**: the rate of enzyme action can be influenced by the temperature. An incubator can be used to keep the temperature constant. This makes the comparison fair and reliable.
- **Concentration of substrate**: the concentration of substrate should be kept constant for all trials. This makes the comparison fair and reliable.
- **Concentration of enzyme**: the concentration of enzyme should be kept constant for all trials. This makes the comparison fair and reliable.

Dependent variable: Rate of enzyme action

Step one: Potato discs contain an enzyme called catalase. This enzyme converts hydrogen peroxide into water and oxygen gas, as shown in the equation below.

$$H_2O_2 \xrightarrow{\text{Catalase}} H_2O + O_2$$

The rate of enzyme action can be measured by measuring the rate of oxygen production, using the equipment shown in figure 3.8.

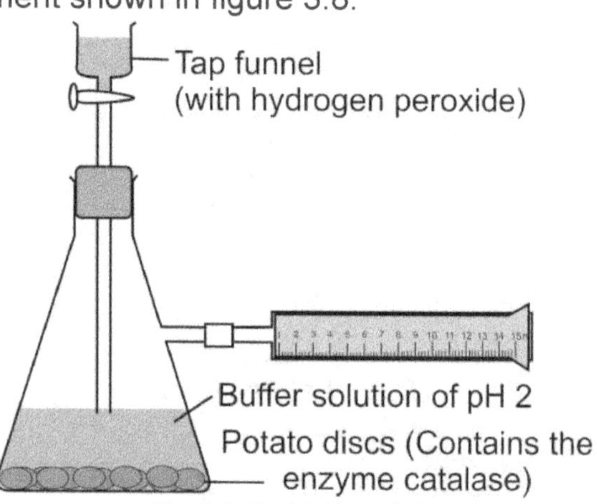

Fig. 3.8: Catalase activity

Step two: Open the tap and allow the hydrogen peroxide to mix with the buffer solution and the potato discs. Oxygen is released. Collect and measure the volume of oxygen in the gas syringe as shown in figure 3.9.

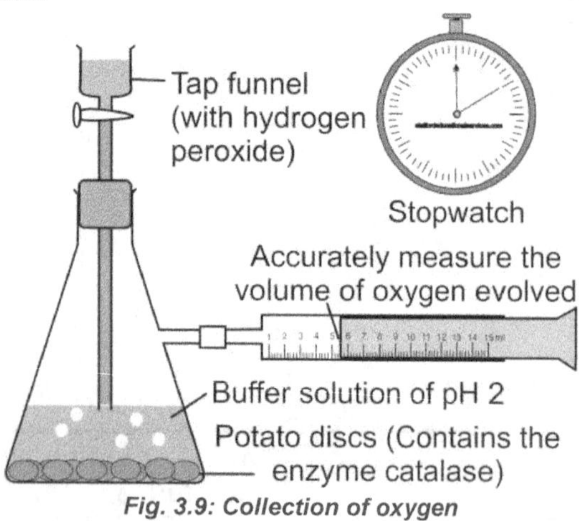

Fig. 3.9: Collection of oxygen

Step three: Measure the rate of reaction.

$$Rate\ of\ reaction = \frac{Volume\ of\ oxygen\ released}{Time\ taken}$$

Step four: Repeat the experiment by using buffer solutions of pH 3, 4, 5, 6, 7, 8, 9, 10, 11 and 12.

The number of potato discs, the size of discs, the volume of buffer solution, the concentration and volume of hydrogen peroxide and temperature must be kept constant for all trials.

Step five: Plot a graph to analyse the results.

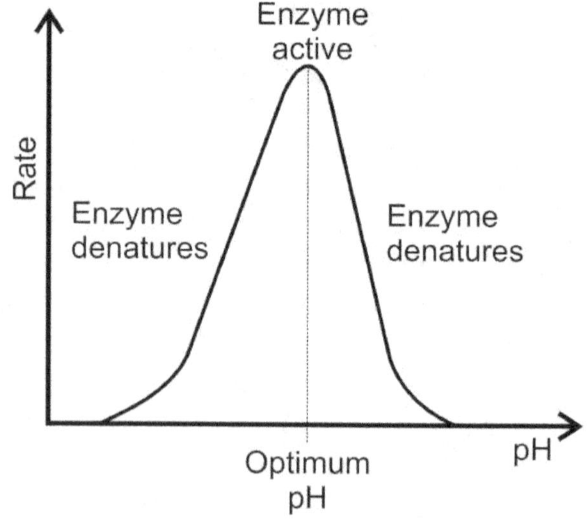

The equation below summarises the process of photosynthesis.

$$\text{Carbon dioxide + Water} \xrightarrow[\text{Light energy}]{\text{Chlorophyll}} \text{Glucose + Oxygen}$$

$$6CO_2 + 6H_2O \xrightarrow[\text{Light energy}]{\text{Chlorophyll}} C_6H_{12}O_6 + 6O_2$$

Raw materials: Carbon dioxide and glucose.
Products: Glucose and Oxygen. The **glucose can then be used to make other organic compounds**.
Role of chlorophyll: Chlorophyll traps light energy and converts it into chemical energy (Glucose).

Testing a leaf for starch

Step one: Boil the leaf in water to stop all metabolic reactions. This prevents starch from being used up during the experiment.

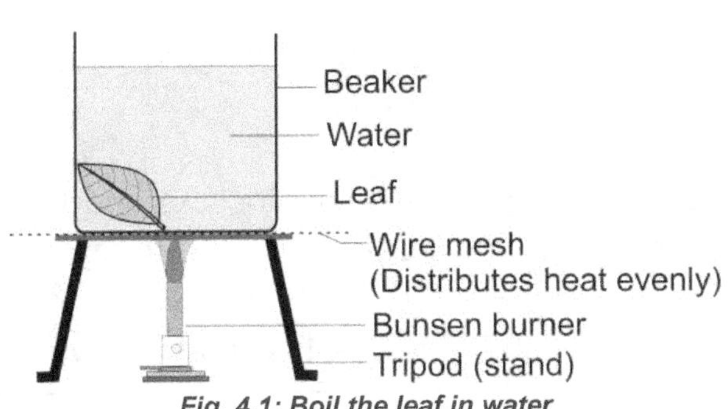

Fig. 4.1: Boil the leaf in water

Step two: Boil the leaf in ethanol to remove chlorophyll. This allows the colour change of iodine to be observed easily. A water bath is used as a safety precaution as ethanol is highly flammable.

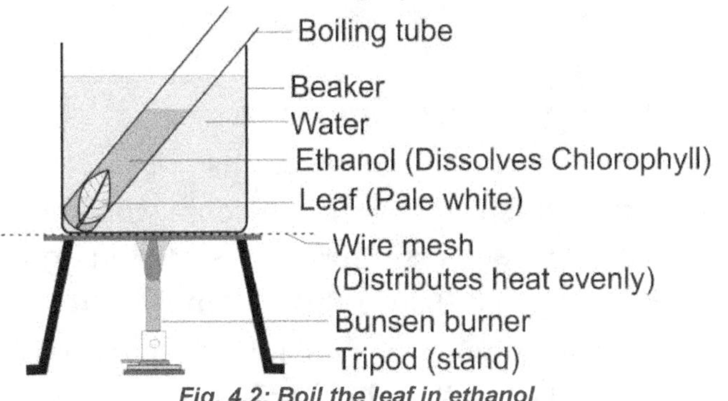

Fig. 4.2: Boil the leaf in ethanol

Step three: Boil the leaf again in water to soften the tissues. The ethanol made the leaf brittle.

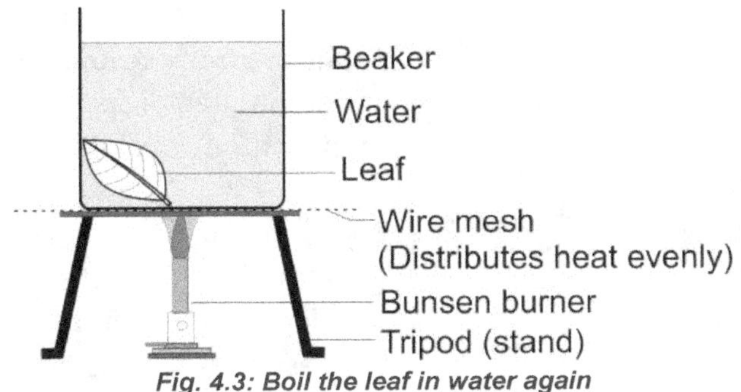

Fig. 4.3: Boil the leaf in water again

Step four: Place the leaf on a white tile and add a few drops of iodine solution. If the iodine changes from **brown to blue-black**, then it confirms that starch is present in the leaf.

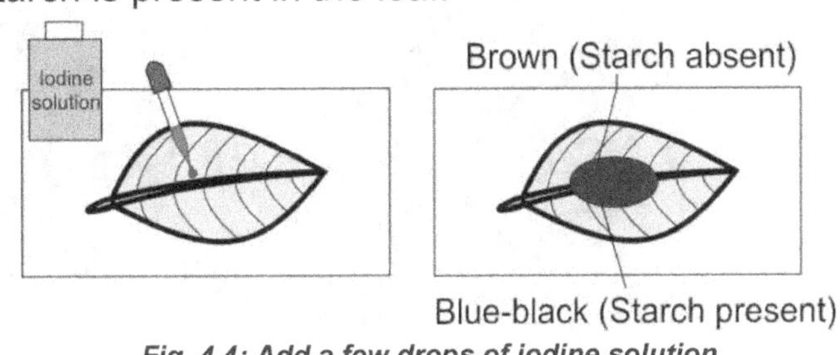

Fig. 4.4: Add a few drops of iodine solution

Note: De-starching a plant before using if for an experiment for photosynthesis will ensure that any starch present in the leaves, at the end of the experiment, was produced during the experimental procedure. This is because there was no starch in the leaves at the start of the experiment.

Experiment to investigate the necessity of chlorophyll for photosynthesis

Step one: De-starch a plant with variegated leaves

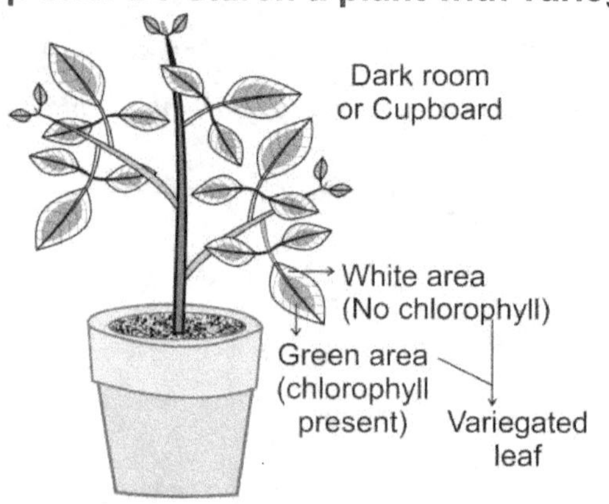

Fig. 4.5: De-starching a plant

- A potted plant, with variegated leaves, can be de-starched by placing it in a dark room or cupboard for 48 hours.
- During this period the plant cannot photosynthesise and will use up all the stored starch for its metabolism.
- A single leaf can be plucked off from the plant and **tested for starch** to confirm whether the plant is completely de-starched.
- If starch is present then the plant can be kept in the dark for a few more hours until all the starch is used up.

Step two: Allow the plant to photosynthesise

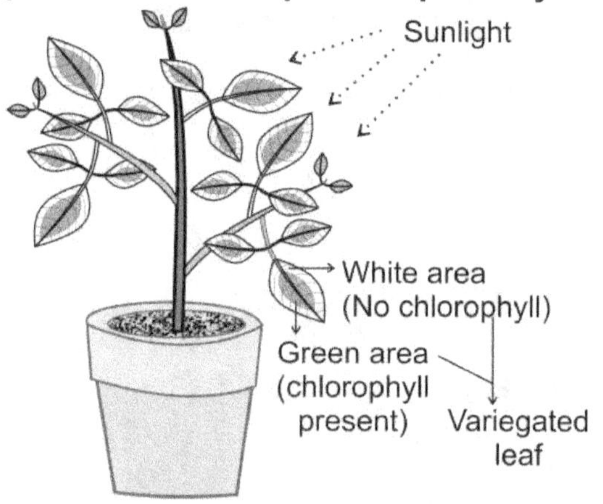

Fig. 4.6: Allow the plant to photosynthesise

- Place the de-starched plant with variegated leaves in sunlight for 30 minutes. The plant photosynthesises during this period and produces glucose.
- The glucose is then converted into starch and stored in the leaf.
- One leaf is then plucked from the plant and tested for starch.

Step three: Interpretation of results

- The non-chlorophyllous regions (white areas) of the leaves does not photosynthesise and the iodine solution remains brown, as shown in figure 4.7. This is due to the absence of chlorophyll in these regions. This is the **experimental setup**.
- The chlorophyllous area photosynthesises. So, the iodine solution turns blue-black, as shown in figure 4.7. This is the **control setup**.

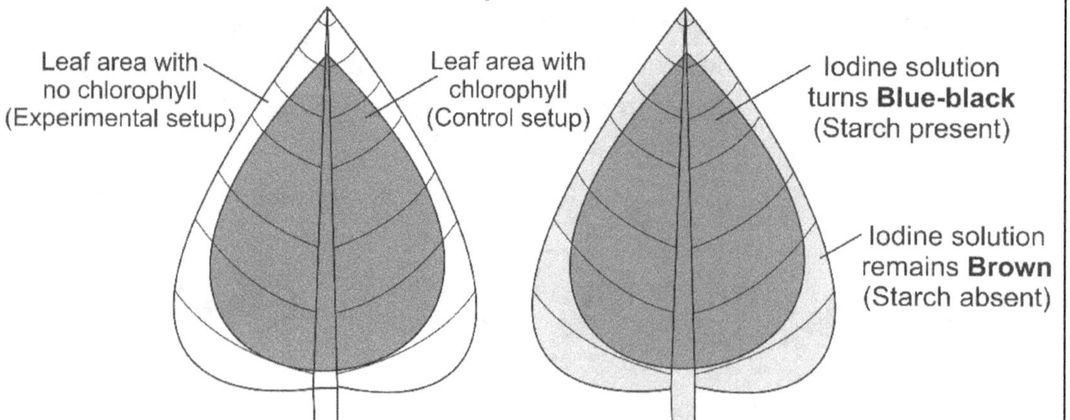

Fig. 4.7: Comparison of control and experimental setup

> **Control setup:** Has all the necessary conditions for photosynthesis.
>
> **Experimental setup:** One condition (chlorophyll) for photosynthesis is lacking.
>
> The results are then compared to make a **valid conclusion**. The difference in the results is because of the missing factor (Chlorophyll).

Conclusion: Chlorophyll is necessary for photosynthesis.

Experiment to investigate the necessity of light for photosynthesis

Step one: De-starch a plant

- A potted plant can be de-starched by placing it in a dark room or cupboard for 48 hours. During this period the plant cannot photosynthesise and will use up all the stored starch for its metabolism.
- A single leaf can be plucked off from the plant and **tested for starch** to confirm whether the plant is completely de-starched.
- If starch is present then the plant can be kept in the dark for a few more hours until all the starch is used up.

Fig. 4.8: De-starch the plant

Step two: Cover one strip of a leaf with aluminium foil

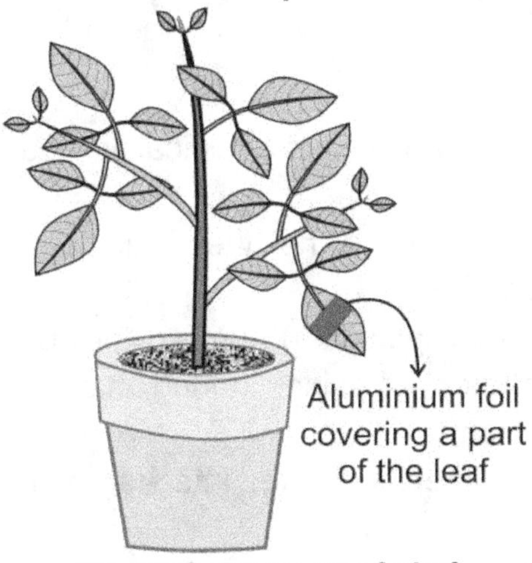

- Cover a part of one leaf with an aluminium foil, as shown in figure 4.9. This ensures that the covered area of the leaf does not receive light.
- The plant is then placed in sunlight for 30 minutes. The plant photosynthesises during this period and produces glucose.
- The glucose is then converted into starch and stored in the leaf.
- The leaf is then plucked from the plant and tested for starch.

Fig. 4.9: Cover an area of a leaf

Step three: Interpretation of results

- The area of the leaf covered by the foil (experimental setup) does not photosynthesise as it does not receive light. Hence there is no starch and the iodine solution remains brown, as shown in figure 4.10.
- The uncovered area of the leaf (control setup) receives light and photosynthesises. Starch is present and the iodine solution turns blue-black, as shown in figure 4.10.

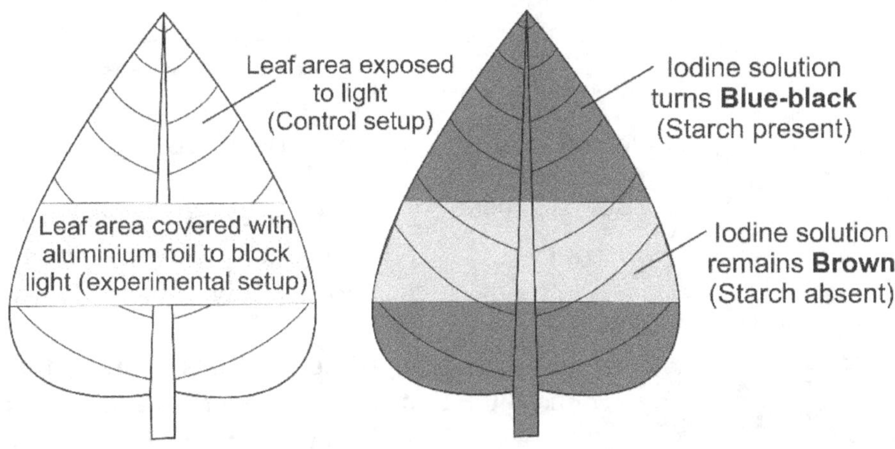

Fig. 4.10: Comparison of control and experimental setup

> **Control setup**: Has all the necessary conditions for photosynthesis.
>
> **Experimental setup:** One condition (light energy) for photosynthesis is lacking.
>
> The results are then compared to make a **valid conclusion**. The difference in the results is because of the missing factor (Light).

Conclusion: Light is necessary for photosynthesis.

Experiment to investigate the necessity of carbon dioxide for photosynthesis

Step one: De-starch a plant

Fig. 4.11: De-starch the plant

- A potted plant can be de-starched by placing it in a dark room or cupboard for 48 hours. During this period the plant cannot photosynthesise and will use up all the stored starch for its metabolism.
- A single leaf can be plucked off from the plant and **tested for starch** to confirm whether the plant is completely de-starched.
- If starch is present then the plant can be kept in the dark for a few more hours until all the starch is used up.

Step two: Cover one strip of a leaf with aluminium foil

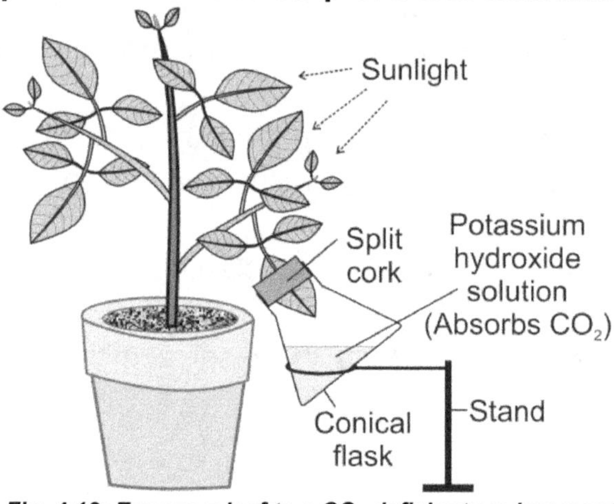

Fig. 4.12: Expose a leaf to a CO₂ deficient environment

- Insert one of the leaves into a conical flask, as shown in figure 4.12.
- The area of the leaf within the flask does not get carbon dioxide.
- The plant is then placed in sunlight for 30 minutes. The plant photosynthesises during this period and produces glucose.
- The glucose is then converted into starch and stored in the leaf.
- The leaf is then plucked from the plant and tested for starch.

Step three: Interpretation of results

- The area of the leaf within the flask (experimental setup) does not photosynthesise and the iodine solution remains brown, as shown in figure 4.13.
- The area of the leaf which receives carbon dioxide (control setup) photosynthesises. So, the iodine solution turns blue-black due to the presence of starch, as shown in figure 4.13.

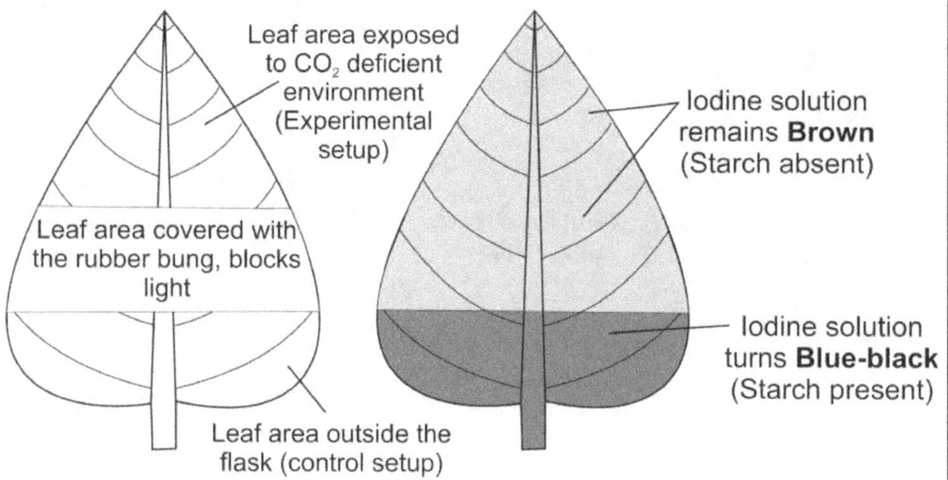

Fig. 4.13: Comparison of control and experimental setup

Control setup: Has all the necessary conditions for photosynthesis.

Experimental setup: One condition (carbon dioxide) for photosynthesis is lacking.

The results are then compared to make a **valid conclusion**. The difference in the results is because of the missing factor (carbon dioxide).

Conclusion: Carbon dioxide is necessary for photosynthesis.

Cambridge 5090 syllabus specification 4(d) investigate and state the effect of varying light intensity, carbon dioxide concentration and temperature on the rate of photosynthesis (e.g. in submerged aquatic plants);

Experiment to investigate the effect of varying light intensity on the rate of photosynthesis

A simple apparatus is setup as shown in figure 4.14 to measure the rate of photosynthesis in submerged aquatic plants, such as Canadian pond weed (*elodea*).

The effect of varying light intensity on the rate of photosynthesis can be investigated as explained below.

Independent variable (The factor which is varied): Light intensity
- The experiment is carried out in a dark room. The only source of light is the electric bulb.
- The light intensity is controlled by using a bulb of the same power (100W).
- Keep the light bulb at a fixed distance from the plant to keep light intensity constant.
- Measure the rate of photosynthesis at different light intensities by moving the bulb further away from the plant. As distance between the bulb and plant increases, the light intensity decreases.

Confounding or controlled variables: Factors that can influence the dependent variable. These must be kept constant to obtain **reliable results**.

- Temperature must be kept constant throughout the experiment by using a water bath at a constant temperature, as shown in figure 4.14.

- Carbon dioxide concentration is kept constant by using a sodium hydrogencarbonate solution of the same concentration throughout the experiment, as shown in figure 4.14. Ideally carbon dioxide should be in excess.

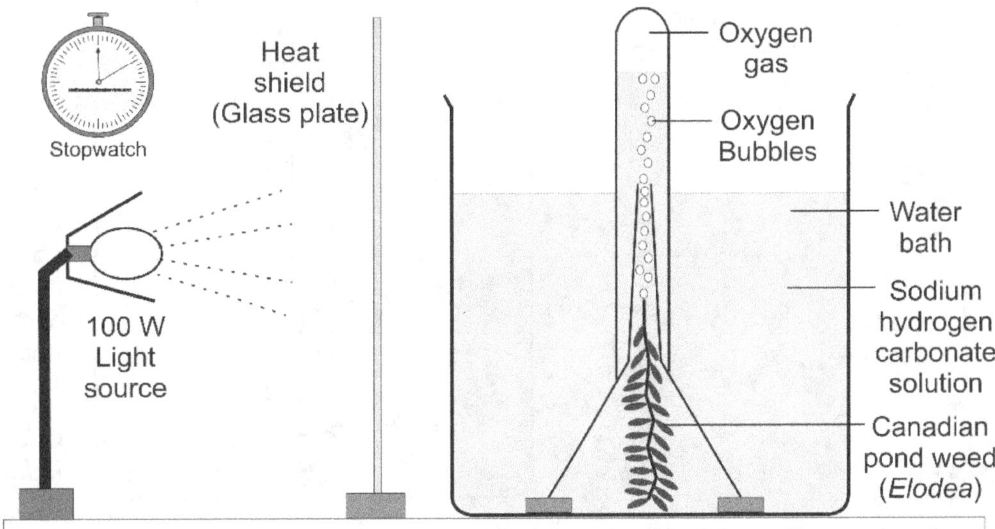

Fig. 4.14: Measuring the rate of photosynthesis in a submerged aquatic plant

Dependent variable (The factor which is measured towards the end of the experiment to provide data for interpretation): rate of photosynthesis.

- Count the number of **oxygen bubbles** released in 2 minutes.

$$Rate\ of\ photosyntehsis = \frac{Number\ of\ bubbles}{Time}$$

Presentation and interpretation of results

The light intensity is plotted on X axis and the rate of photosynthesis is plotted on the Y axis of a line graph. The trends and patterns on the graph are used for the interpretation of results.

Investigating the effect of varying carbon dioxide concentration on the rate of photosynthesis
The apparatus in figure 4.14 counts the number of oxygen bubbles evolved during photosynthesis. This gives an **inaccurate** measure of photosynthesis as the volume of each bubble may differ. Moreover the bubbles are released as a continuous stream and will be difficult to count accurately.

The apparatus can be modified by using a measuring cylinder and setup as shown in figure 4.15.

The effect of varying carbon dioxide concentration on the rate of photosynthesis can be investigated as explained below.

Independent variable (The factor which is varied): Carbon dioxide concentration
Measure the rate of photosynthesis at different carbon dioxide concentrations by varying the concentration of sodium hydrogen carbonate in the solution.

Confounding or controlled variables: Factors that can influence the dependent variable. These must be kept constant to obtain **reliable results**.
Light intensity
- The experiment is carried out in a dark room. The only source of light is the electric bulb.
- The light intensity is controlled by using a bulb of the same power (100W).
- Keep the light bulb at a fixed distance from the plant to keep light intensity constant.

Temperature
Temperature must be kept constant throughout the experiment by using a water bath at a constant temperature, as shown in figure 4.15.

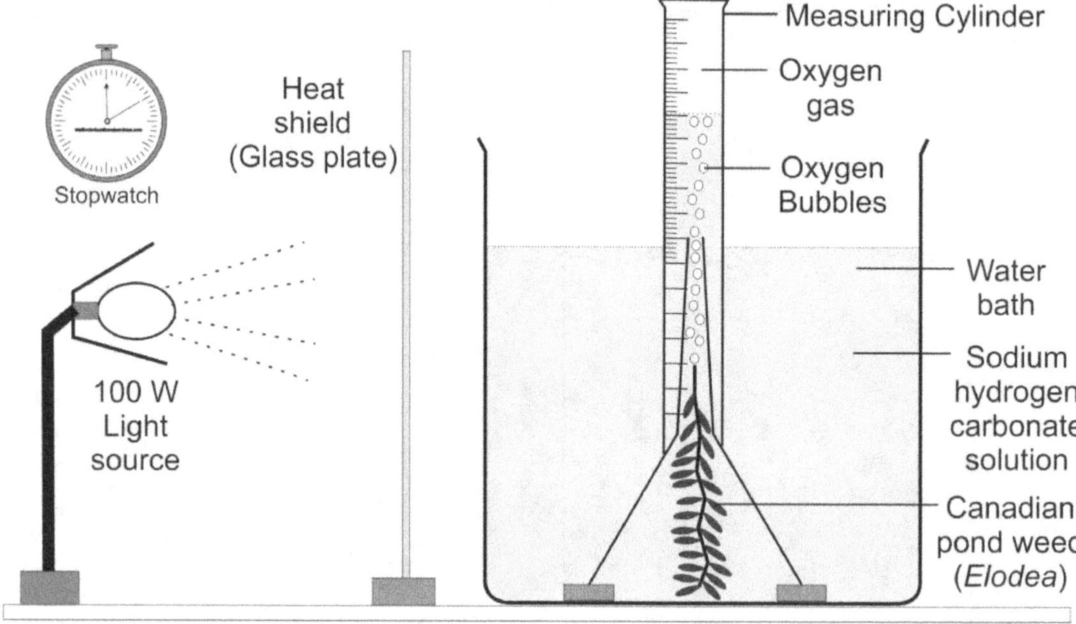

Fig. 4.15: Measuring the rate of photosynthesis in a submerged aquatic plant

Dependent variable (The factor which is measured towards the end of the experiment to provide data for interpretation): rate of photosynthesis.

$$\text{Rate of photosynthesis} = \frac{Volume\ of\ oxygen\ evolved}{Time}$$

The volume of oxygen is measured in the measuring cylinder and a stopwatch is used to measure time.

Presentation and interpretation of results
The carbon dioxide concentration is plotted on X axis and the rate of photosynthesis is plotted on the Y axis of a line graph. The trends and patterns on the graph are used for the interpretation of results.

Investigating the effect of varying temperature on the rate of photosynthesis
The effect of varying temperature on the rate of photosynthesis can be investigated as explained below.

Independent variable (The factor which is varied): Temperature
Measure the rate of photosynthesis at different temperatures by varying the temperature of the water bath.

Confounding or controlled variables: Factors that can influence the dependent variable. These must be kept constant to obtain **reliable results**.

Light intensity
- The experiment is carried out in a dark room. The only source of light is the electric bulb.
- The light intensity is controlled by using a bulb of the same power (100W).
- Keep the light bulb at a fixed distance from the plant to keep light intensity constant.

Carbon dioxide concentration
Carbon dioxide concentration is kept constant by using a sodium hydrogencarbonate solution of the same concentration throughout the experiment, as shown in figure 4.16. Ideally carbon dioxide should be in excess.

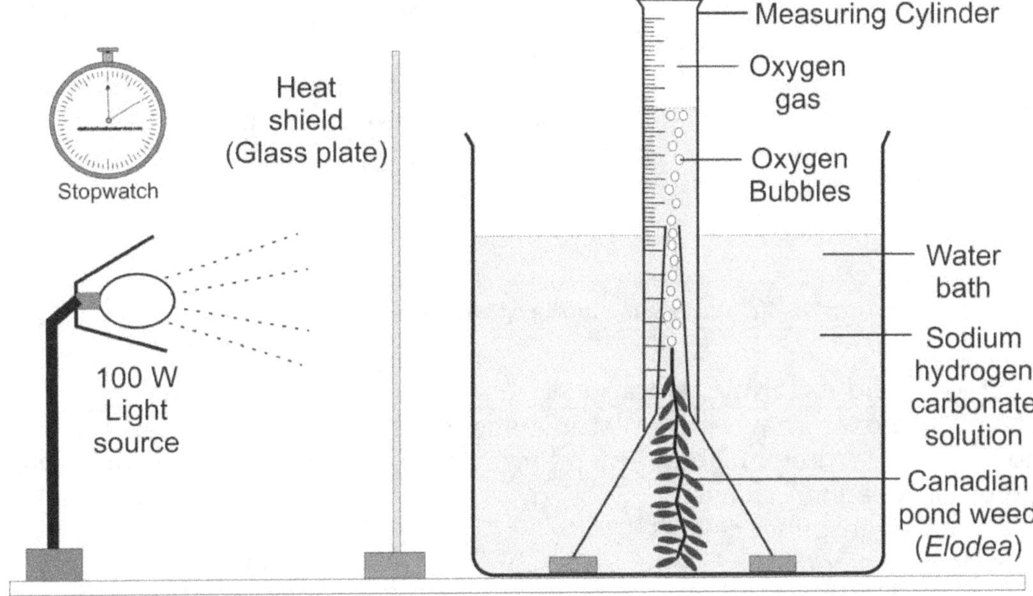

Fig. 4.16: Measuring the rate of photosynthesis in a submerged aquatic plant

Dependent variable (The factor which is measured at towards the end of the experiment to provide data for interpretation): rate of photosynthesis.

$$\text{Rate of photosynthesis} = \frac{Volume\ of\ oxygen\ evolved}{Time}$$

The volume of oxygen is measured in the measuring cylinder and a stopwatch is used to measure time.

Presentation and interpretation of results
The carbon dioxide concentration is plotted on X axis and the rate of photosynthesis is plotted on the Y axis of a line graph. The trends and patterns on the graph are used for the interpretation of results, as explained in figure 4.17.

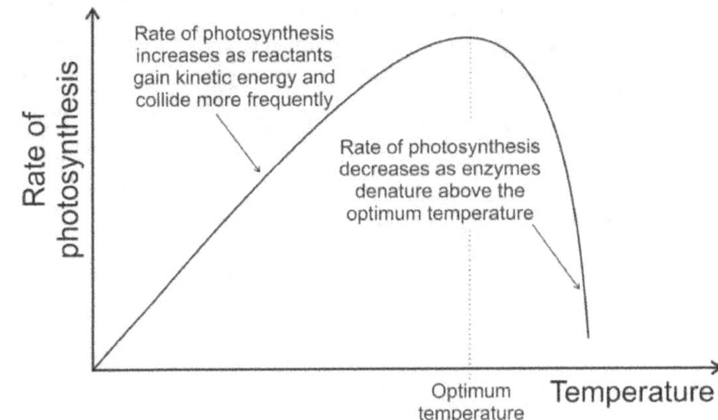

Fig. 4.17: Effect of temperature on the rate of photosynthesis

Law of limiting factors: When a process is influenced by several factors, the rate at which the process proceeds is determined by the factor in the **shortest supply**.

Photosynthesis is influenced by light intensity, temperature and carbon dioxide concentration. At zero light intensity (in darkness) the rate of photosynthesis is zero. As light intensity increases the rate of photosynthesis also increases, but only up to a certain point. At this point the CO_2 concentration or temperature, must be limiting the rate of photosynthesis, as shown in figure 4.18.

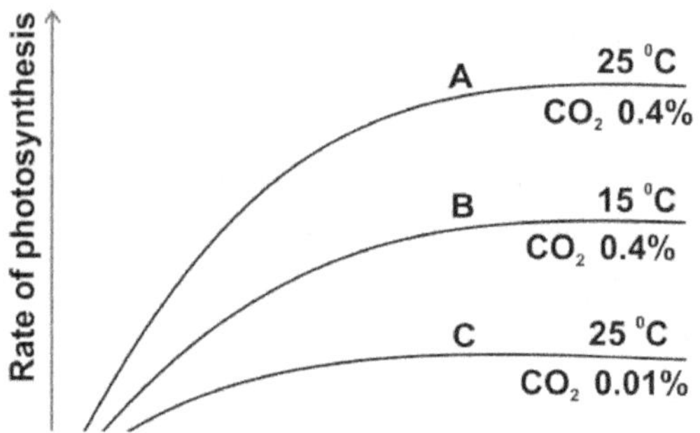

Fig. 4.18: Limiting factors and light saturation point

- The rate of photosynthesis is **lower in curve C than in curve A**, even though temperature is optimum. This is because the **CO_2 concentration of 0.01% is the limiting factor** in experiment C.
- The rate of photosynthesis is lower in curve B than in curve A, even though CO_2 concentration is optimum. This is because the low temperature of 15 ^0C **is the limiting factor** in experiment B.
- In curve **A**, the rate becomes constant because it has reached the **light saturation point**. This is the maximum rate of photosynthesis possible by the plant.

Intake of carbon dioxide and water by plants

Carbon dioxide diffuses from the atmosphere into the leaf through the stomata. The carbon dioxide then diffuses into the chloroplasts of the mesophyll cells, where it is used for photosynthesis.	**Water** is absorbed by the roots of plants. The pathway followed by water is shown in the flowchart below.

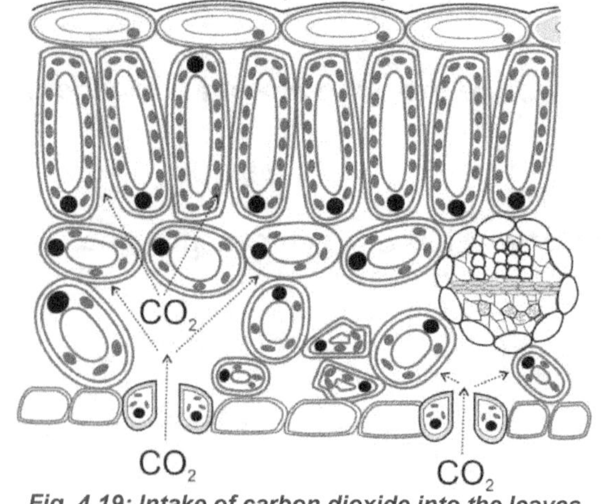

Fig. 4.19: Intake of carbon dioxide into the leaves

Water in soil
↓ **Osmosis**
Water in root hair cells
↓ **Diffusion and osmosis**
Water in xylem vessels of roots
↓ **Root pressure**
Transpiration pull
Capillary action
Water in xylem vessels of leaves → Water in mesophyll cells used for photosynthesis
↓ **Evaporation**
Water vapour in air spaces of mesophyll layer
↓ **Diffusion**
Water vapour in atmosphere

Autotrophs are organisms which prepare their own food, usually by photosynthesis. Chlorophyll in plants is used to trap the light energy into chemical energy in the form of glucose. The glucose is then used to synthesise large complex organic molecules like starch, proteins and lipids.

Heterotrophs cannot prepare their own food. They obtain energy from the breakdown of complex readymade organic molecules, obtained from the environment. Heterotrophic organisms may be herbivores or carnivores, which directly or indirectly obtain energy from autotrophs.

Figure 4.20 shows the structure of a dicotyledonous leaf as seen under a microscope. The functions of the various parts is described below.

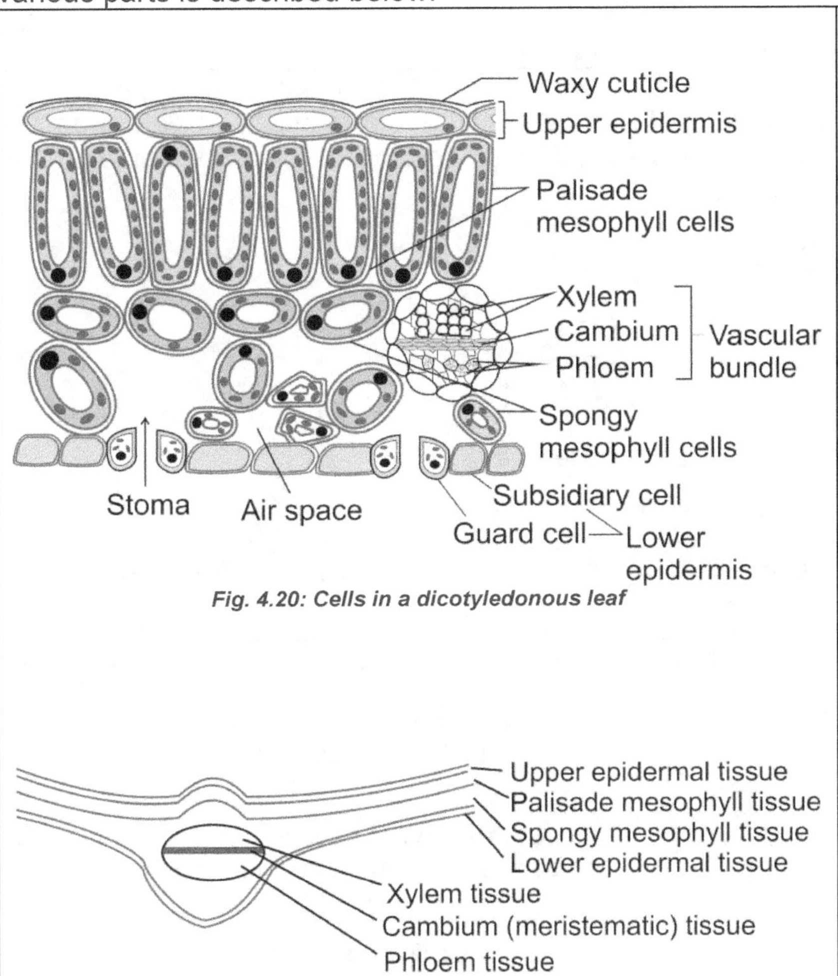

Fig. 4.20: Cells in a dicotyledonous leaf

Fig. 4.21: Plan diagram showing position of tissues in a dicotyledonous leaf

- The **waxy cuticle** is transparent and waterproof. It allows light penetration but reduces transpiration.

- The **upper epidermis** is a protective layer which allows light penetration and prevents entry of pathogens.

- The **palisade mesophyll cells** have the **most chloroplasts**. They are cylindrical to absorb maximum light. **Air spaces** between the cells also **increases the surface area** for absorption of carbon dioxide. The large vacuole pushes the chloroplasts towards the cell membrane and ensures that carbon dioxide does not have to travel too far to reach the chloroplasts after diffusing into the cells.

- The **spongy mesophyll cells** have **many chloroplasts**. They are irregular in shape to form **air spaces** between the cells. This **increases the surface area** for absorption of carbon dioxide.

- **Guard cells** are used for opening of stomata to allow transpiration and absorption of carbon dioxide. The guard cells have a **few chloroplasts** to enable the opening and closing of stomata.

- Xylem vessels transport water and ions into the leaf.

- Phloem carries the prepared food away from the leaf.

Opening and closing of stomata

Opening of Stomata	Closing of Stomata
<ul><li>Solutes in guard cells increase and accumulate.</li><li>Water potential of guard cells decreases.</li><li>Water moves into the guard cells from subsidiary cells by osmosis.</li><li>Guard cells become turgid and more curved, due to unequal elasticity of cell wall, as shown in figure 4.22.</li><li>Stomata opens.</li></ul>	<ul><li>Solutes in guard cells decrease in concentration.</li><li>Water potential of guard cells increases.</li><li>Water moves from the guard cells into subsidiary cells by osmosis.</li><li>Guard cells become flaccid and less curved, due to unequal elasticity of cell wall, as shown in figure 4.22.</li><li>Stomata closes.</li></ul>

Fig. 4.22: Opening and closing of stomata

Wilting refers to the loss of turgidity in soft tissue of plants. This occurs when the rate of loss of water from the plant is greater than the absorption of water by the plant. It is a mechanism to prevent excess water loss. The stomata also close during wilting to reduce transpiration.

Fig. 4.23: Wilting of leaves

Cambridge 5090 syllabus specification 4(j) understand the effect of a lack of nitrate and magnesium ions on plant growth.

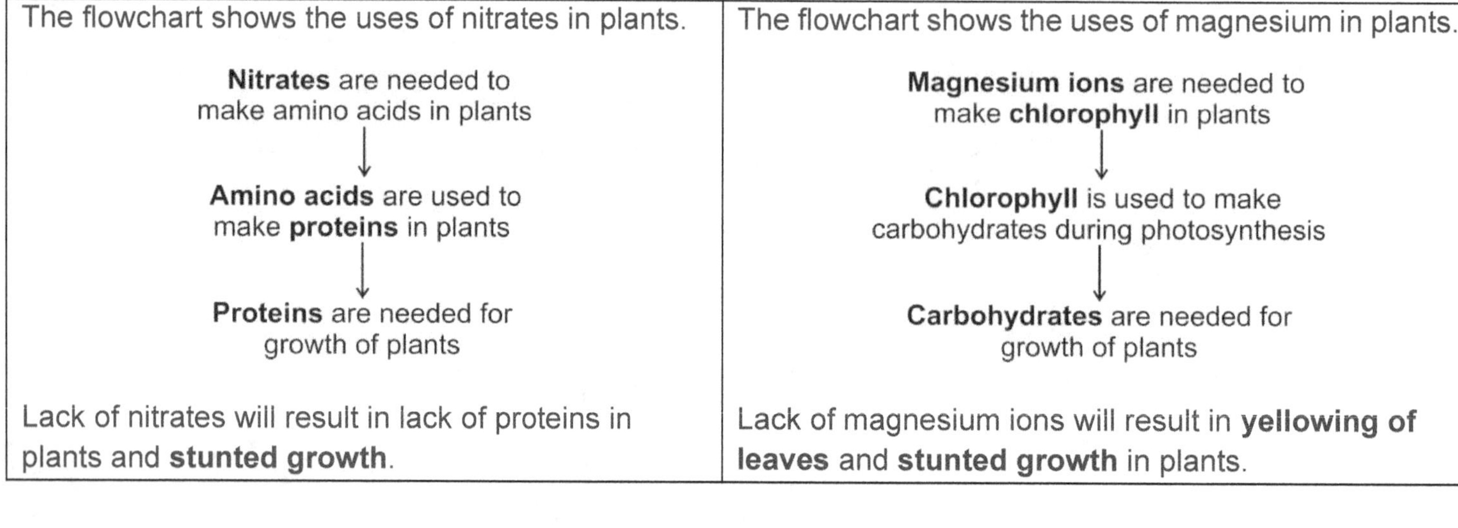

The flowchart shows the uses of nitrates in plants.	The flowchart shows the uses of magnesium in plants.
Nitrates are needed to make amino acids in plants ↓ **Amino acids** are used to make **proteins** in plants ↓ **Proteins** are needed for growth of plants Lack of nitrates will result in lack of proteins in plants and **stunted growth**.	**Magnesium ions** are needed to make **chlorophyll** in plants ↓ **Chlorophyll** is used to make carbohydrates during photosynthesis ↓ **Carbohydrates** are needed for growth of plants Lack of magnesium ions will result in **yellowing of leaves** and **stunted growth** in plants.

Cambridge 5090 syllabus specification 5(a) *list the chemical elements that make up: carbohydrates; fats; proteins;*
Cambridge 5090 syllabus specification 5(b) *describe tests for: starch (iodine in potassium iodide solution); reducing sugars (Benedict's solution); protein (biuret test); fats (ethanol emulsion test);*

The major components of food are carbohydrates, fats and proteins. The table below lists the chemical elements found in food.

Chemical compound	Chemical elements
Carbohydrates	Carbon, Hydrogen, Oxygen only (C,H,O)
Fats (lipids)	Carbon, Hydrogen, Oxygen only (C,H,O)
Proteins	Carbon, Hydrogen, Oxygen, Nitrogen (C,H,O,N). Sulphur may also be present

Test for starch

Starch is insoluble in water. To test food for the presence of starch, iodine in potassium iodide solution can be added directly to the food. If the iodine changes colour from brown to **blue-black,** it confirms the presence of starch, as shown in figure 5.1.

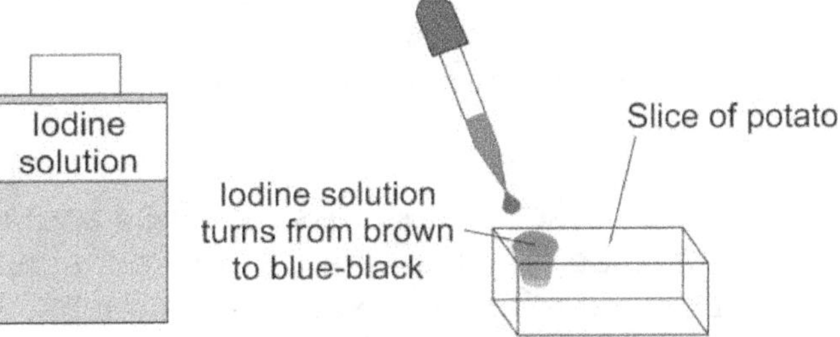

Fig. 5.1: Testing for starch

Benedict's test for reducing sugars

- Weigh 50 g of the food and blend it in a food blender in 200 cm³ of distilled water to form a pulp.
- Filter the pulp and collect the filtrate. The residue can be discarded.

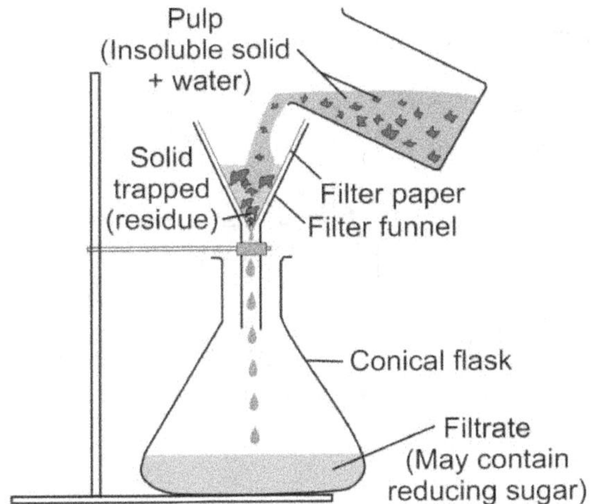

Fig. 5.2: Preparing a food extract

Note: *If the food to be tested is **already liquefied**, such as milk or juice then the **preparation of extract is not required**. Benedict's test can directly be carried out as shown in figure 5.3.*

- Heat 5cm³ of the filtrate with 5 5cm³ of Benedict's solution in a water bath, as shown in figure 5.3.

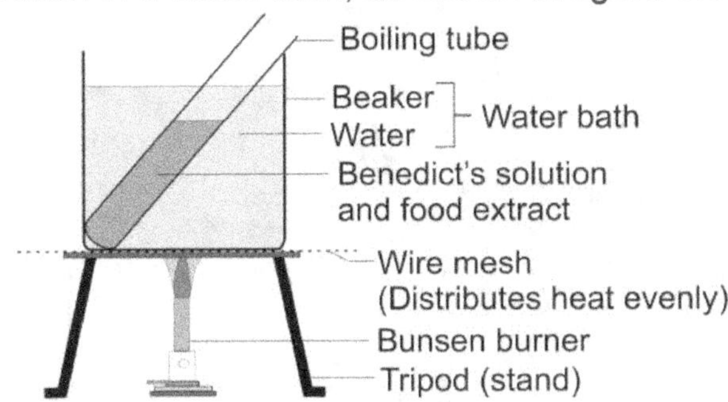

Fig. 5.3: Benedict's test

If the colour of the Benedict's solution changes from **blue to green to yellow to orange to red**, then reducing sugar is present. **Green and yellow indicates low concentration** of reducing sugar. **Orange and brick red indicate high concentration** of reducing sugar. If **two solutions** form a red precipitate then filter and dry the red precipitate. The solution which produces higher mass of red precipitate has a higher concentration of reducing sugar.

The **water bath** allows gradual heating to prevent spurting of liquids from the test tube. This is a **safety precaution**.

Biuret test for proteins

If the food is solidified (meat or soya), then prepare a food extract as explained in figure 5.2. If the food is liquefied, such as egg albumin or milk, then preparation of extract is not required.

Add Biuret solution to the food extract of liquid food. If the Biuret solution changes colour from blue to violet or lilac, then it confirms the presence of protein, as shown in figure 5.4.

The intensity of the purple colour is proportionate to the protein concentration in the food.

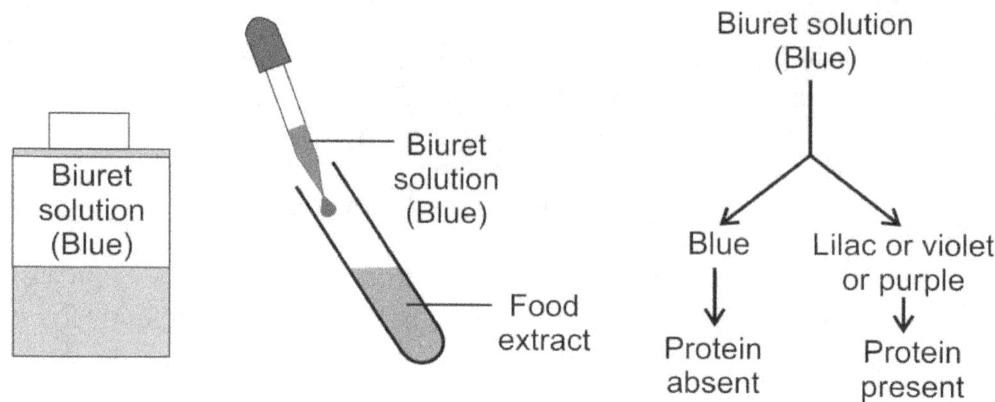

Fig. 5.4: Biuret test for proteins

Ethanol emulsion test for Lipids

If the food sample is **solid**, like groundnuts or soya, then crush the food sample by using a mortar and pestle. Transfer the crushed sample to a dry test tube and add ethanol. Shake the mixture thoroughly. Allow the solid to settle down for a few minutes and pour the ethanol into another test tube. This is the food extract. Perform the ethanol emulsion test as shown in figure 5.5.

If the food sample is already a liquid, like coconut oil, then directly carry out the ethanol emulsion test as shown in figure 5.5.

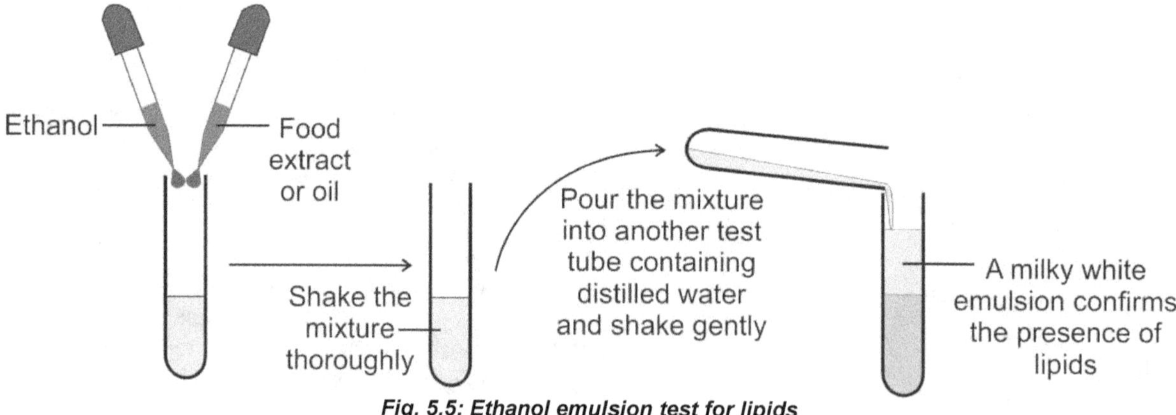

Fig. 5.5: Ethanol emulsion test for lipids

- A milky white emulsion confirms the presence of lipids.
- A clear colourless mixture confirms the absence of lipids.

Dietary component	Principal source	Dietary importance	Deficiency symptom
Carbohydrates	Rice and potatoes	• Glucose is used as a respiratory substrate to provide energy for metabolic reactions, like nerve impulse transmission and cell division. • Cellulose prevents constipation.	Tiredness
Fats	Full cream milk, cheese, beef, butter	• Provides energy for metabolism. • Serves as a solvent for vitamins. • Serves as a thermal insulator for people living in cold climates.	Tiredness
Proteins	Lean meat, egg white and soya.	• Used to make proteins essential for growth and body building. • Used to make enzymes, insulin, haemoglobin, antibodies, etc.	Muscle wasting
Vitamin C	Citrus fruits (Oranges, Lemons)	• Needed for wound healing.	**Scurvy**: anaemia, bleeding spots on the skin, loose teeth, bleeding gums.
Vitamin D	Milk, cheese	• Helps in the uptake and storage of calcium and phosphorus, which keeps bones and teeth strong and healthy.	**Rickets**: soft bones, resulting in weak and deformed bandy legs.
Calcium	Milk, cheese	• Needed for strong bones and teeth.	Bandy legs
Iron	Beef, Liver, Tuna	• Needed to make haemoglobin, which carries oxygen to the cells.	**Anaemia (Low haemoglobin levels): Fatigue**
Fibre (roughage)	Vegetables and fruits	• Helps the food to pass easily through the alimentary canal. This prevents constipation and lowers the risk of bowel cancer.	**Constipation, increased risk of bowel cancer**
Water	Fruits, vegetables and drinking water	• A solvent for transport of glucose in the blood and removal of urea. • A medium for chemical reactions and enzyme action. • As a coolant during sweating.	**Fatigue, Overheating of the body (Hyperthermia)**

Balanced diet

Food comprises of nutrients and chemical energy. A balanced diet should contain the exact amount of energy the body needs for metabolism and movement. It should also contain all the essential nutrients in the right proportion. An imbalanced diet that lacks the essential nutrients, or, which has too much or too little energy compared to the needs of the body will lead to malnutrition. The amount of energy and nutrients needed by the body may vary from person to person. Figure 5.6 summarises the requirements of diet with respect to age, gender and activity levels.

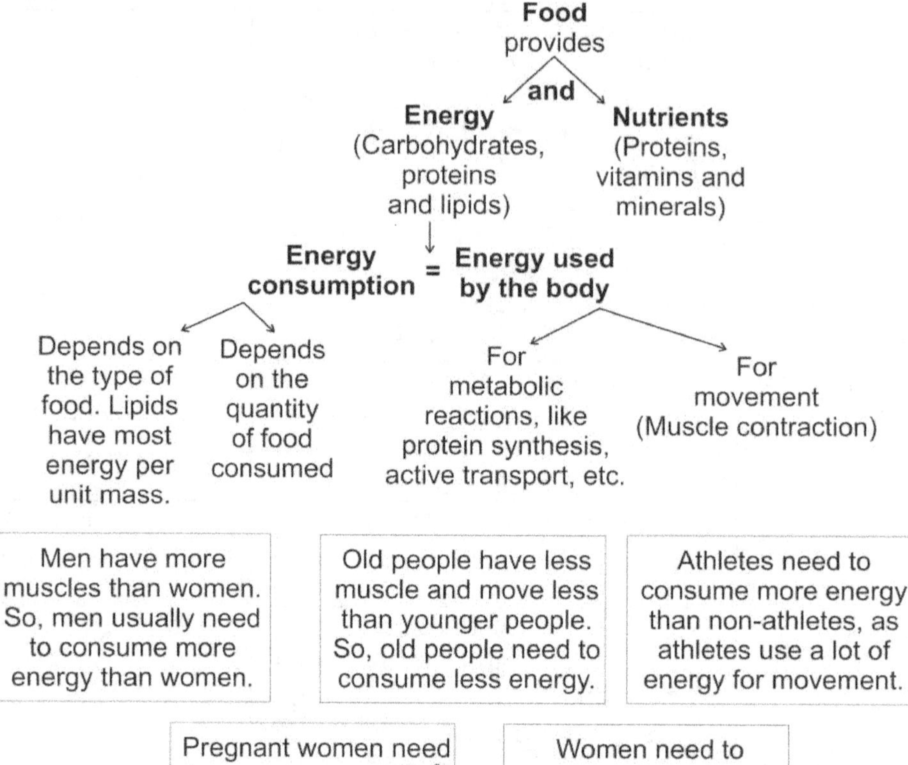

Fig. 5.6: Dietary requirements

Cambridge 5090 syllabus specification 5(g) state the effects of malnutrition in relation to starvation, heart disease, constipation and obesity;
Cambridge 5090 syllabus specification 5(h) discuss the problems that contribute to famine (unequal distribution of food, drought and flooding, increasing population);

Malnutrition is a term which refers to both undernourishment and over-nourishment.

Undernourishment can result when the body does not receive the required amount of energy that is needed for metabolism and physical activity. It can result in starvation and deficiency diseases.	**Over-nourishment** can result when the body receives more energy than it requires for metabolism and physical activity. It results in obesity and heart diseases.

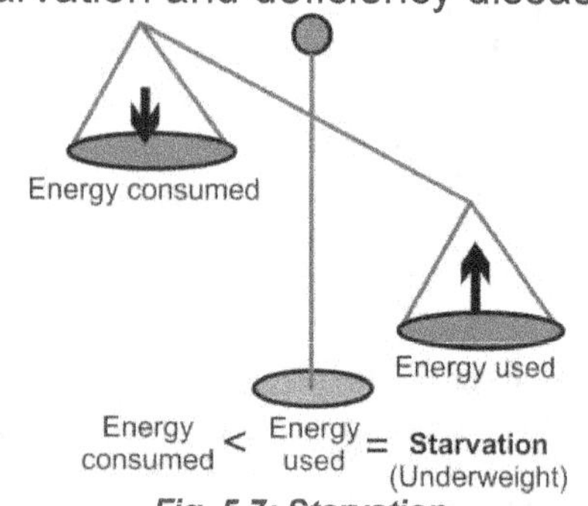

Fig. 5.7: Starvation

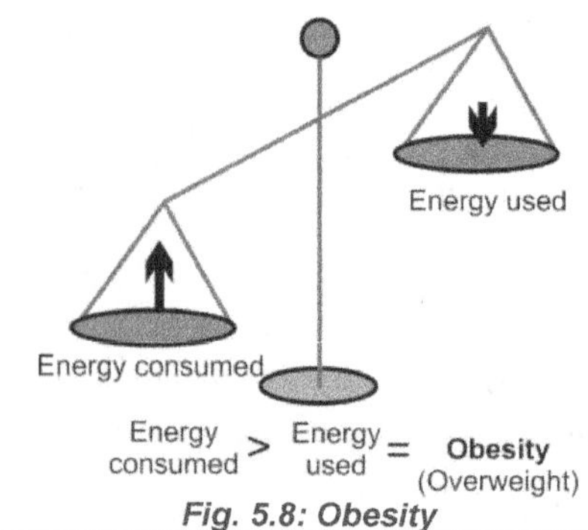

Fig. 5.8: Obesity

Starvation	Obesity
Starvation is often the result of famine. Famine is the lack of food to support a population. Starvation leads to the depletion of glycogen reserves, followed by the depletion of fat reserves in adipose tissue. Finally the body starts to obtain energy from the breakdown of proteins in skeletal muscles. This severely weakens the individual.	Obesity is one of the major preventable risk factors of **heart diseases.** Obesity leads to high blood pressure. This increases the risk of formation of fatty deposits (plaque or atheroma) in the coronary arteries. The blocking of coronary arteries by plaque can result in heart attacks.
Famine may result due to: <ul><li>Poverty</li><li>Overpopulation</li><li>Drought</li><li>Flooding</li><li>War or political instability</li><li>Crop failure due to pests or unfavourable climatic conditions.</li><li>Growing of cash crops instead of food.</li></ul>	**Constipation** is a condition in which there is difficulty in emptying the bowels, usually associated with hardened faeces. It is caused due to lack of fibre in the diet. Constipation can be avoided by consuming plenty of fruits and vegetables.

Cambridge 5090 syllabus specification 5(i) identify the main regions of the alimentary canal and the associated organs: mouth (buccal) cavity, salivary glands, oesophagus, stomach, duodenum, pancreas, gall bladder, liver, ileum, colon, rectum and anus;
Cambridge 5090 syllabus specification 5(j) describe the main functions of these parts in relation to ingestion, digestion, absorption, assimilation and egestion of food, as appropriate;
Cambridge 5090 syllabus specification 5(m) describe peristalsis;
Cambridge 5090 syllabus specification 5(o) describe:
• digestion in the alimentary canal;
• the functions of a typical amylase, protease and lipase, listing the substrates and end-products;

Figure 5.9 shows the different glands and regions of the alimentary canal. It is necessary to be able to identify all the parts labelled on the diagram.

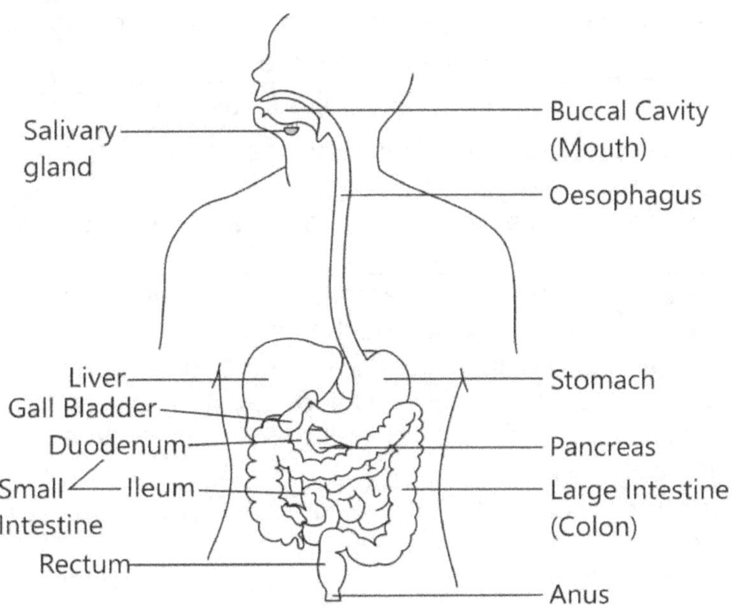

Fig. 5.9: Human alimentary canal and glands

Structure	Functions
Mouth (buccal) cavity **(Ingestion and digestion)**	• Mastication (chewing) of food increases the surface area for enzyme action and speeds up digestion. The teeth play a vital role in crushing and grinding of food. • Food is mixed with saliva and formed into a **bolus**. The slippery mucus makes it easy to swallow the food bolus. • Digestion of starch into maltose by salivary amylase. $$\text{Starch} \xrightarrow[\text{pH 7 to 7.5}]{\text{Salivary amylase (Ptyalin)}} \textbf{Maltose}$$
Salivary glands **(Digestion)**	• Secretes salivary amylase and mucus in saliva.
Oesophagus **(Digestion)**	• Moves the food bolus from the mouth to the stomach by peristalsis. • **Peristalsis** is the rhythmic contraction and relaxation of the circular and longitudinal muscles of the oesophagus which propels the food through the alimentary canal. 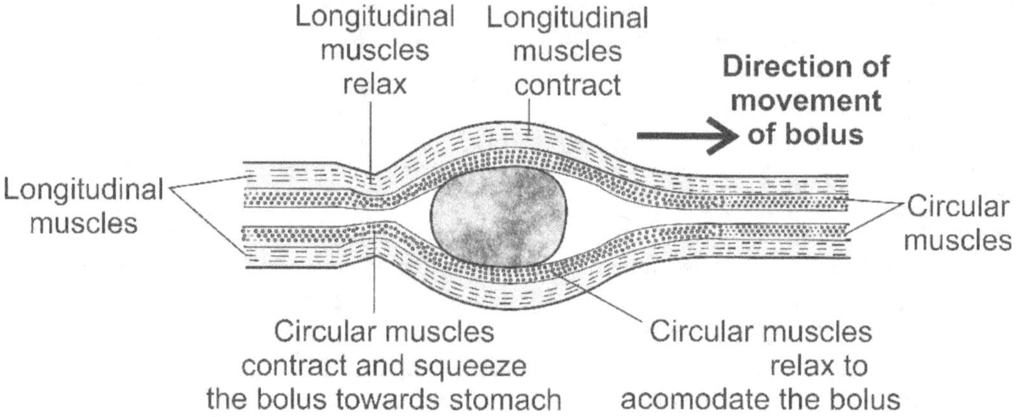 Fig. 5.10: Peristalsis in the oesophagus • The digestion of starch continues in the oesophagus until the food reaches the stomach and pH changes. $$\text{Starch} \xrightarrow[\text{pH 7 to 7.5}]{\text{Salivary amylase (Ptyalin)}} \textbf{Maltose}$$
Stomach **(Digestion)**	• Pepsin is an enzymes found in the gastric juice of the stomach. $$\textbf{Protein} \xrightarrow[\text{pH 3 (HCl in gastric juice)}]{\text{Pepsin (protease)}} \textbf{Polypeptides and dipeptides}$$ • Hydrochloric acid is also found in the stomach and maintains an optimum pH for pepsin. • The Hydrochloric acid also kills bacteria that may be present in the food.
Pancreas **(Digestion)**	• The pancreas produces pancreatic juice, which is delivered to the duodenum by the pancreatic duct. • The pancreatic juice contains bicarbonate ions which neutralise the Hydrochloric acid from the stomach and maintain an alkaline pH for the pancreatic enzymes.
Gall bladder **(Digestion)**	• Bile is produced by the liver and stored in the Gall bladder. • The bile salts also neutralise the Hydrochloric acid from the stomach and maintain an alkaline pH for the pancreatic enzymes. • Bile also emulsifies fats (lipids) to increase the surface area for the action of lipase.

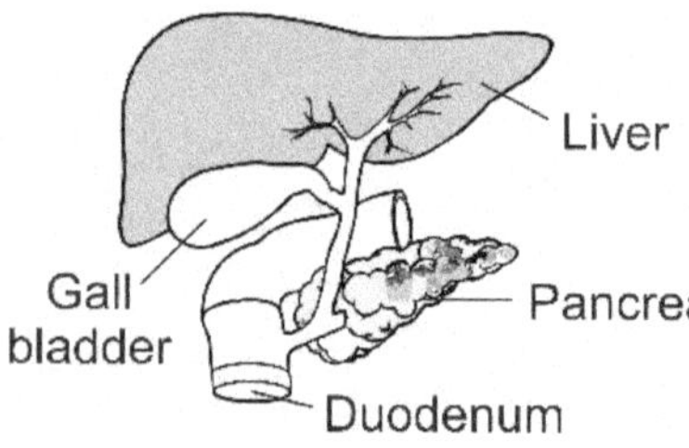

Fig. 5.11: Liver and pancreas

Duodenum **(Digestion)**	• The duodenum is the place where the food from the stomach enters the small intestine and mixes with pancreatic enzymes and bile.
	• The pancreatic enzymes and their action are illustrated below.
	    - Pancreatic amylase,
	    - Pancreatic lipase and
	    - Pancreatic protease (trypsin)
	
	$$\text{Polypeptides} \xrightarrow[\substack{\text{pH 8 (Bicarbonates neutralise}\\ \text{hydrochloric acid from the stomach)}}]{\text{Trypsin (protease) in pancreatic juice}} \text{dipeptides}$$
	$$\text{Lipids} \xrightarrow[\text{pH 7 to 7.5}]{\text{Pancreatic lipase}} \text{Fatty acids and glycerol}$$
	$$\text{Starch} \xrightarrow[\text{pH 7 to 7.5}]{\text{Pancreatic amylase}} \text{Maltose}$$

Liver **(Digestion and assimilation)**	• The liver produces bile and stores it in the gall bladder. The bile is delivered to the duodenum through the bile duct. • The liver also plays an important role in assimilation of food. The digested food is taken to the liver through the hepatic portal vein. • In the liver, the excess amino acids are de-aminated to form urea. $$\text{Excess Amino acids} \xrightarrow[\text{In liver}]{\text{Deamination}} \text{Urea}$$ • The excess glucose is stored as glycogen and stored in the liver. $$\text{Excess Glucose} \xrightarrow[\text{In liver}]{\text{Insulin}} \text{Glycogen}$$

Ileum **(Digestion and absorption)**	• Digestion is finally complete in the ileum by enzymes in the intestinal juice. The intestinal juice is secreted by the wall of the ileum. The reactions and enzymes involved are shown below.
	
	$$\text{Lipids} \xrightarrow[\text{pH 8}]{\text{Intestinal lipase}} \text{Fatty acids and glycerol}$$
	$$\text{Dipeptides} \xrightarrow[\text{pH 8}]{\text{Erepsin (protease) in intestinal juice}} \text{Amino acids}$$
	$$\text{Maltose} \xrightarrow[\text{pH 8}]{\text{Maltase}} \text{Glucose}$$
	
	The final products of digestion which are absorbed into the blood capillaries in the villi of the ileum are:
	    - Glucose
	    - Amino acids
	    - Fatty acids and glycerol

Colon **(Absorption)**	The function of the colon is to absorb water from the gut.
Rectum **(Egestion)**	The rectum stores the faeces temporarily before releasing it to the outside.
Anus **(Egestion)**	The anus has a sphincter muscle which is used to control the release of faeces during defaecation

A tooth has three layers, as shown in figure 5.12

- The hard outer layer is called enamel.
- The middle layer is called dentine.
- The center of the tooth is called the pulp. It contains nerves and blood vessels. The roots are embedded into the jaw bone.

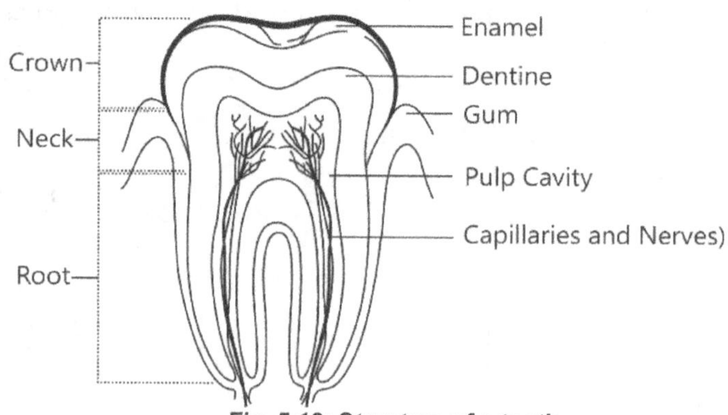

Fig. 5.12: Structure of a tooth

A human adult has 32 teeth, 16 on each jaw. The arrangement of teeth on a single jaw is shown in figure 5.13. The function and structure of the incissors, canines, premolars and molars are also listed in figure 5.13.

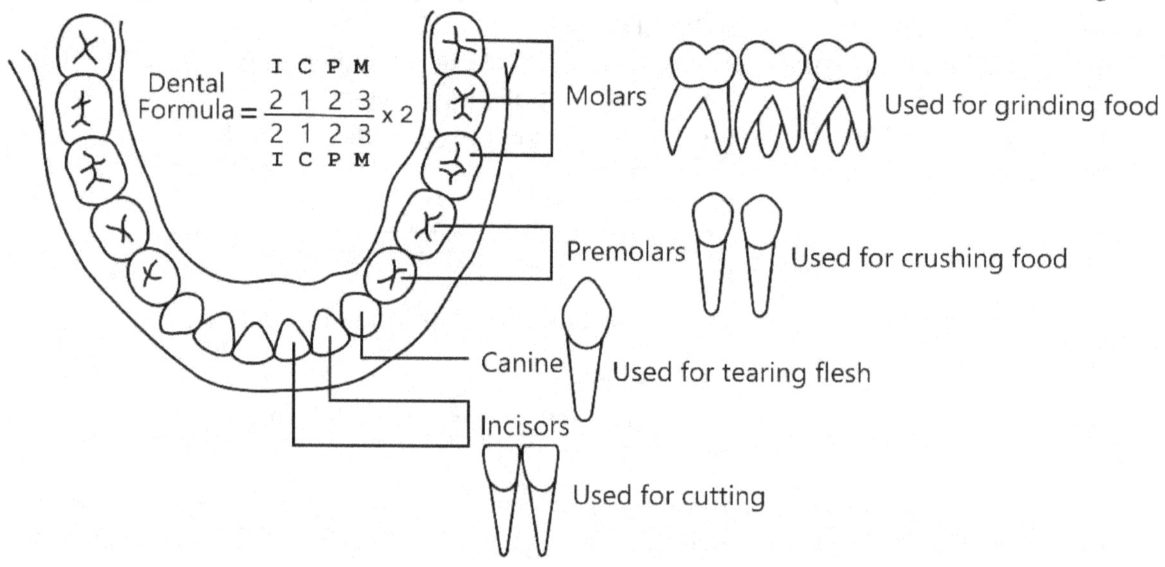

$$\text{Dental Formula} = \frac{2\ 1\ 2\ 3}{2\ 1\ 2\ 3} \times 2$$

Fig. 5.13: Different types of teeth in humans

Tooth decay or dental caries occurs when bacteria in the mouth produce acids that dissolve the enamel and dentine, forming a cavity. If not treated, the bacteria, acids and food can reach the pulp cavity and cause pain, infection, and tooth loss, as shown in figure 5.14.

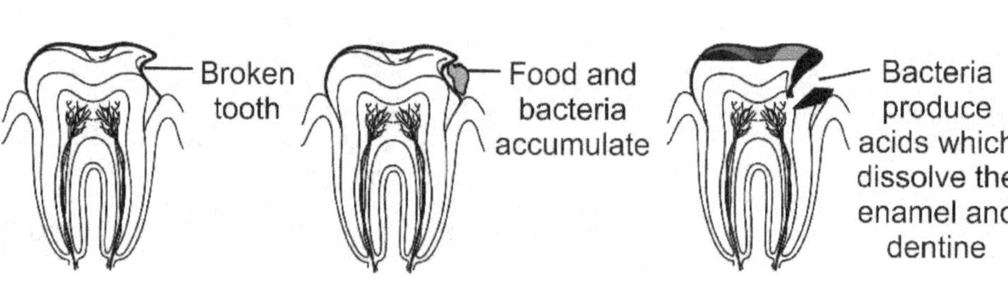

Fig. 5.14: Tooth decay

Tooth decay can be prevented by the following ways.
- Brushing teeth regularly after meals to remove bacteria and food. The alkaline pH of the toothpaste also helps to neutralize acids produced by the bacteria.
- Avoid eating foods that are high in sugar. Bacteria thrive on sugars in the mouth.
- Fluoride helps prevent tooth decay by making teeth more resistant to acids produced by plaque. Fluoride is added to many public water supplies and toothpastes.
- Flossing of teeth after meals helps to remove food particles from between the teeth. This will help to slow bacterial growth and prevents tooth decay.

Undigested food is large and insoluble. It cannot pass from the gut into the blood capillaries. The food is digested to make it soluble and small enough to pass out of the gut wall and enter into the blood capillaries. The blood can then carry the food to the cells, as shown in figure 5.15.

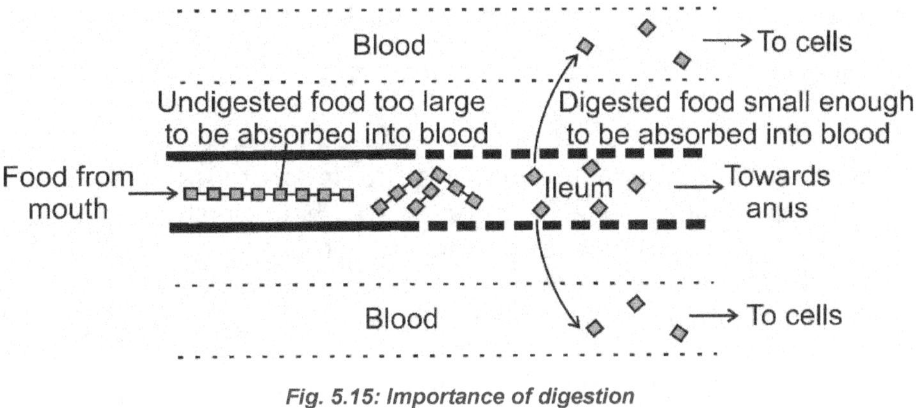

Fig. 5.15: Importance of digestion

Figure 5.16 shows the numerous villi in the ileum which are used to absorb the digested food into the blood and lymph. The large number of villi increase the surface area for absorption.

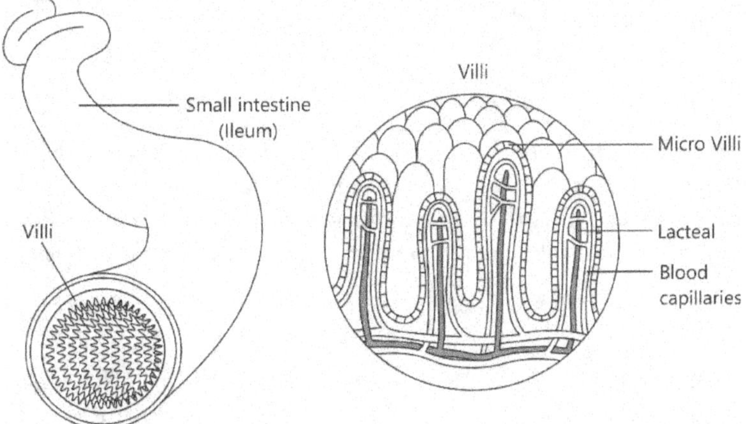

Fig. 5.16: Numerous villi in the ileum increase the surface area for absorption of digested food

The villus is made up of a single layer of epithelial cells to enable rapid absorption of the digested food, as shown in figure 5.17.

The glucose and amino acids are absorbed into the blood capillaries, as shown in figure 5.17.

The fatty acids and glycerol are absorbed into the lacteals, as shown in figure 5.17.

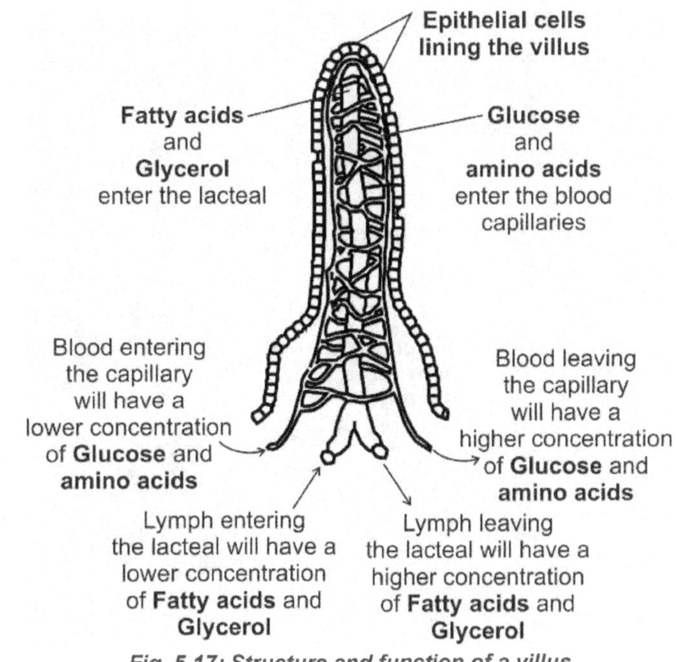

Fig. 5.17: Structure and function of a villus

Chapter Six

Transport in Flowering Plants

Cambridge 5090 syllabus specification 6(a) *relate the structure and functions of root hairs to their surface area and to water and ion uptake;*

Functions of a root hair cell related to its structure

- The root hair cell has a protrusion, root hair, which increases the surface area for absorption of mineral ions, like nitrates, phosphates and magnesium, from the soil by active transport.
- The root hair cell also has a large surface area for the uptake of water by osmosis. Oxygen can also diffuse rapidly into the root hair cell from the soil. The oxygen is essential for respiration, which provides energy for active transport of ions.
- The cell wall of the root hair cell provides mechanical support by becoming turgid.

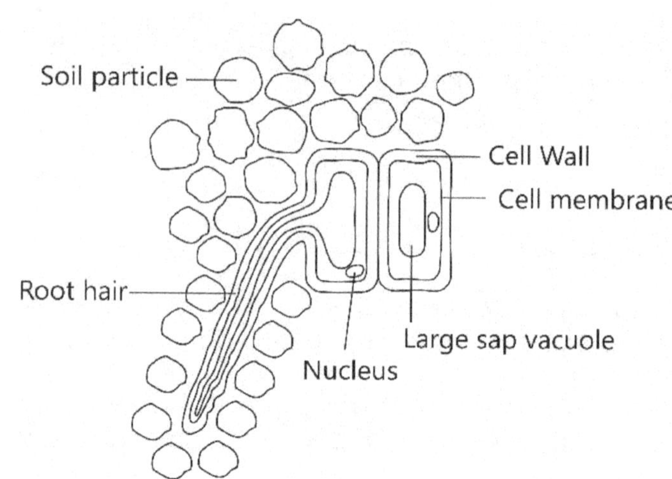

Fig. 6.1: Structure of root hair cell

Cambridge 5090 syllabus specification 6(b) *state that transpiration is the evaporation of water at the surfaces of the mesophyll cells followed by the loss of water vapour from the leaves through the stomata;*
Cambridge 5090 syllabus specification 6(c) *describe: • how water vapour loss is related to cell surfaces, air spaces and stomata;*
• the effects of air currents (wind), and the variation of temperature, humidity and light intensity on transpiration rate;
• how wilting occurs;

Transpiration

Transpiration is the loss of water vapour from the aerial parts of a plant. The water first evaporates from the surface of spongy mesophyll cells and then the water vapour diffuses from the air spaces into the atmosphere, through the stomata, as shown in figure 6.2.

The air spaces increase the surface area of the spongy mesophyll cells and increases the evaporation of water from the surface of the cells.

The water vapour accumulates in the air spaces and then diffuses into the atmosphere down a concentration gradient, through the stomata.

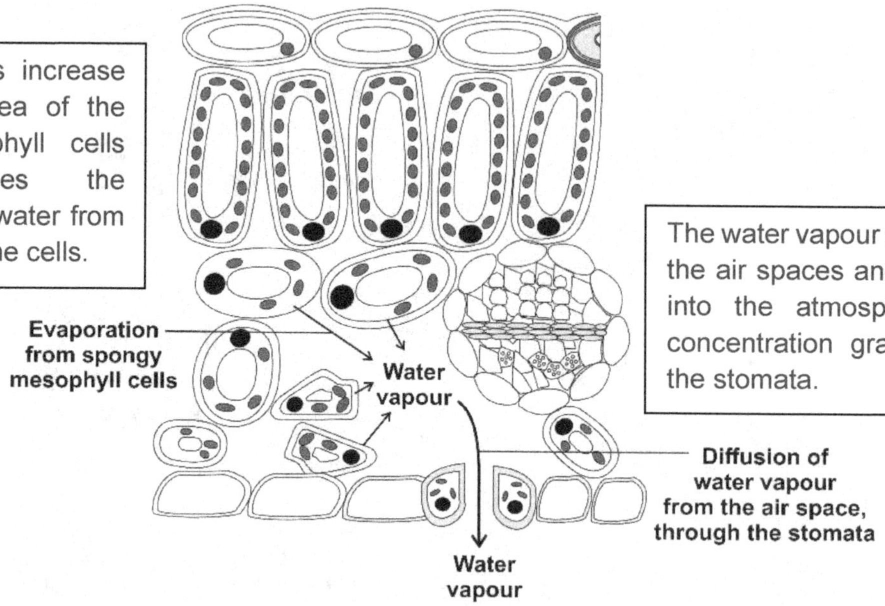

Fig. 6.2: Transpiration

Effect of temperature on transpiration rate

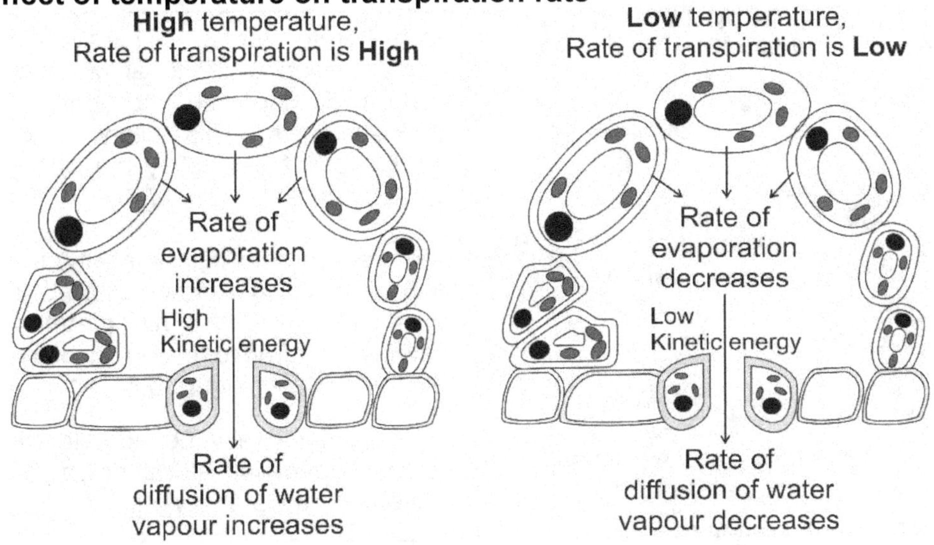

High temperature,
Rate of transpiration is **High**

Low temperature,
Rate of transpiration is **Low**

Fig. 6.3: Effect of temperature on the rate of transpiration

At higher temperature the water molecules have more kinetic energy. The rate of evaporation from mesophyll cells and the rate of diffusion of water vapour increases. So, transpiration rate increases, as shown in figure 6.3.

At lower temperature the water molecules have less kinetic energy. The rate of evaporation from mesophyll cells and the rate of diffusion of water vapour decreases. So, transpiration rate decreases, as shown in figure 6.3.

Effect of humidity on transpiration rate

Low Humidity,
Transpiration rate is **High**

High Humidity,
Transpiration rate is **Low**

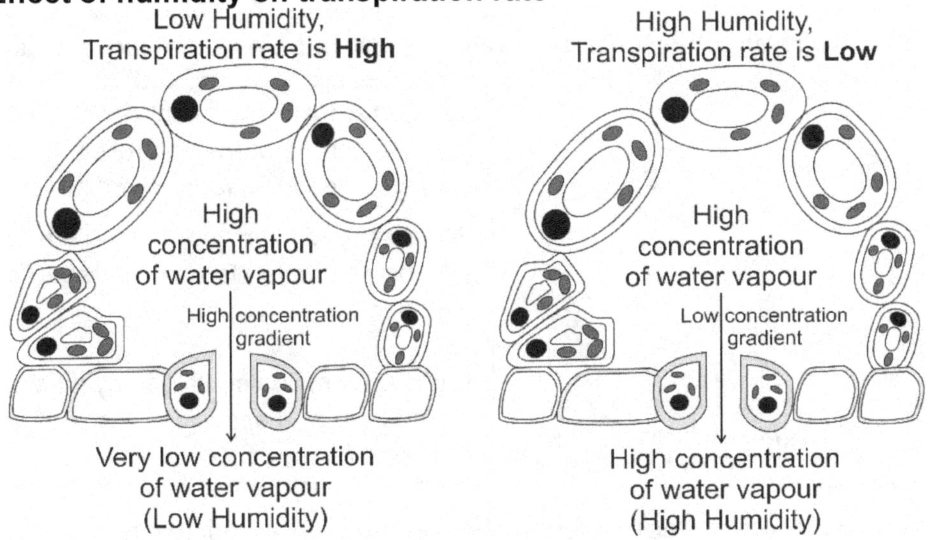

Fig. 6.4: Effect of humidity on the rate of transpiration

When humidity is high then the rate of transpiration is low. This is because the rate of diffusion of water vapour decreases due to the low concentration gradient, as shown in figure 6.4.

When humidity is low then the rate of transpiration is high. This is because the rate of diffusion of water vapour increases due to the high concentration gradient, as shown in figure 6.4

Effect of wind velocity on transpiration rate

High wind velocity,
Transpiration rate is **High**

Low wind velocity,
Transpiration rate is **Low**

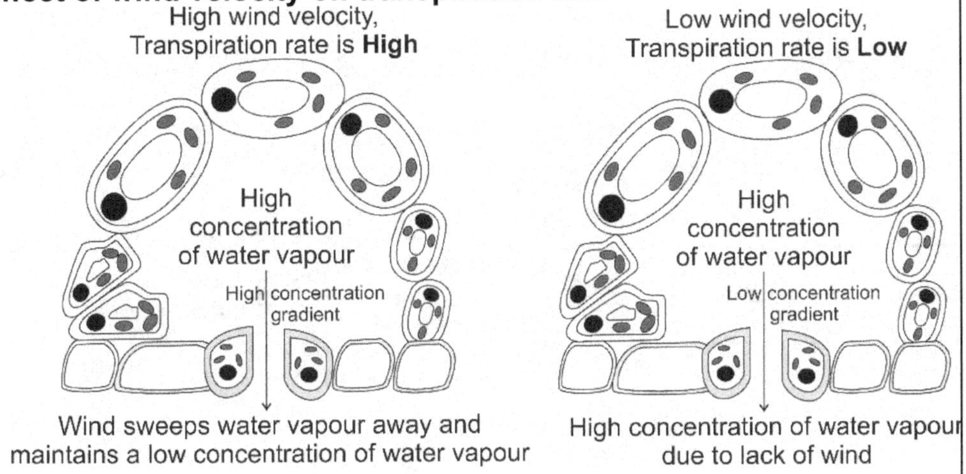

Fig. 6.5: Effect of air currents (wind) on the rate of transpiration

As the wind velocity increases the rate of transpiration increases because the wind sweeps away water molecules as soon as they leave the stomata. This maintains a high concentration gradient and increases the rate of transpiration, as shown in figure 6.5.

As the wind velocity decreases the rate of transpiration decreases because the water molecules accumulate outside the stomata. This maintains a low concentration gradient and decreases the rate of transpiration, as shown in figure 6.5.

Effect of light intensity on transpiration rate

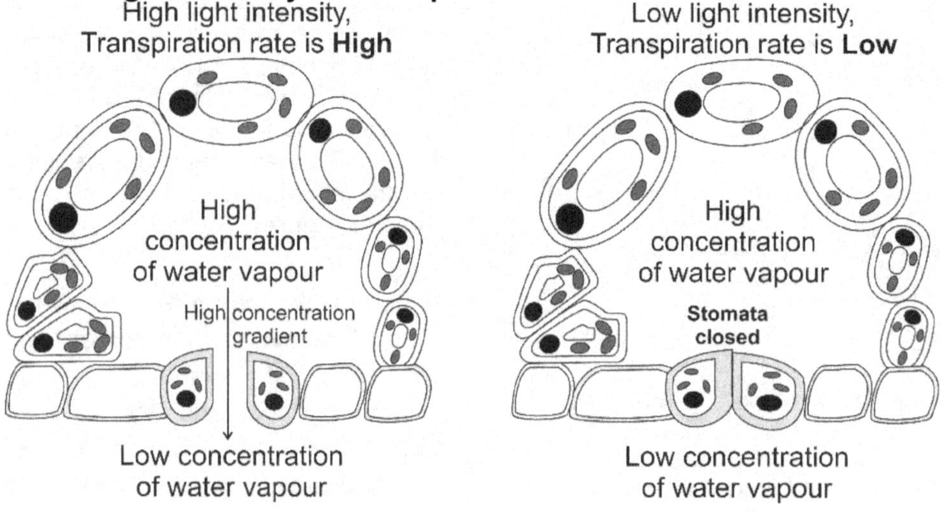

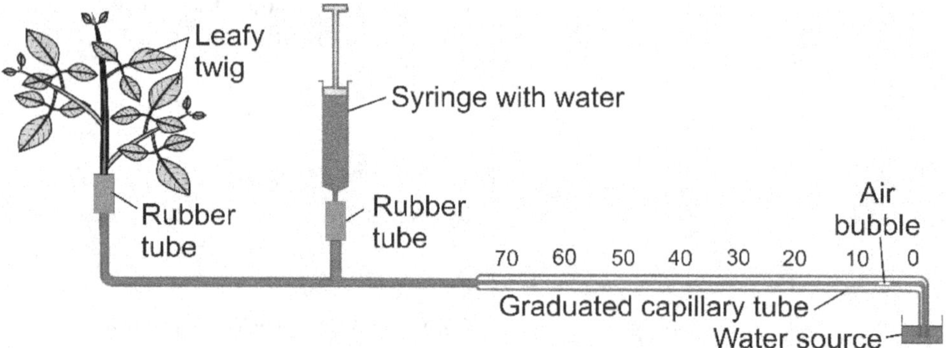

Fig. 6.6: Effect of temperature on the rate of transpiration

As the light intensity increases the stomata open wider to allow more Carbon dioxide into the leaf for photosynthesis. This allow water vapour to diffuse rapidly out of the leaf, as shown in figure 6.6.

In darkness, zero light intensity, the stomata close as carbon dioxide is no longer needed. This prevents water vapour from diffusing out of the leaf, as shown in figure 6.6.

At low light intensity the stomata are partially open and slows down the movement of water vapour.

A potometer can be used to measure the rate of transpiration by a leafy shoot. The apparatus is setup as shown in figure 6.7. The syringe is used to reset the apparatus by pushing the bubble back to the zero mark on the capillary tube.

The rate of transpiration can be measured by recording the distance moved by the bubble in the capillary tube in a fixed interval of time.

Fig. 6.7: A potometer

Temperature can be kept constant by using a temperature controlled room with an air conditioner or room heater.

Wind speed can be controlled by placing a table fan at a specific distance from the apparatus.

Light intensity is controlled by placing a light bulb at a specific distance from the apparatus.

Most transpiration occurs from the lower surface of the leaf, through the stomata. The cuticle on the upper surface reduces water loss as it is waterproof. This can be demonstrated by the following experiment, shown in figure 6.8. The cobalt chloride paper on the lower surface turns pink faster than on the upper surface. This confirms that water vapour is lost more rapidly from the lower (ventral) surface of the leaf.

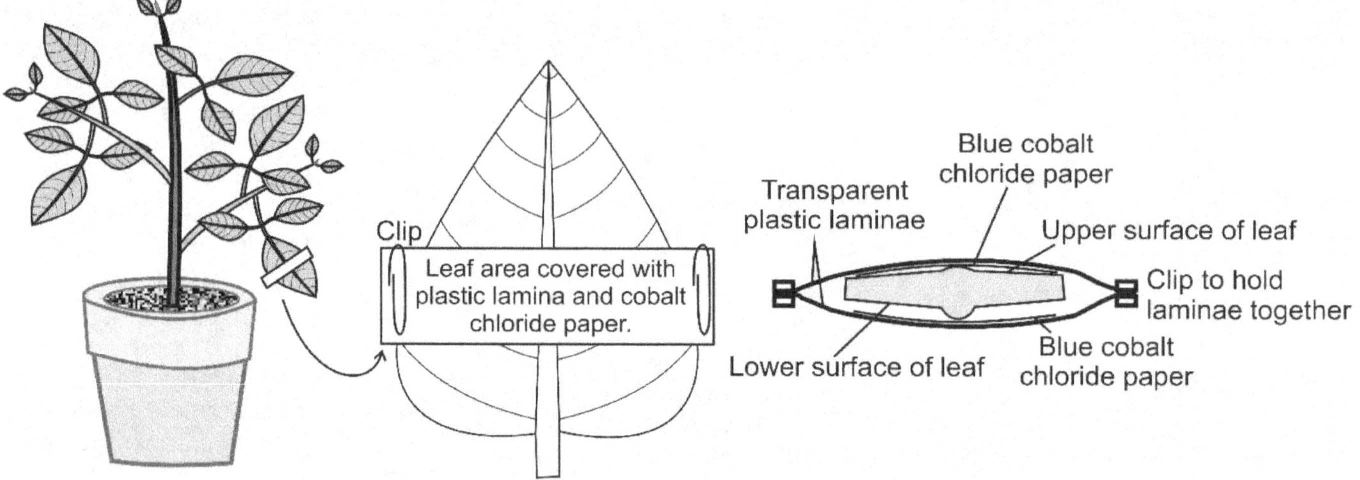

Fig. 6.8: More transpiration occurs from the lower surface of the leaf

Cambridge 5090 syllabus specification 6(d) investigate, using a suitable stain, the pathway of water in a cut stem;

Water and mineral ions are transported from the roots to the leaves through the xylem vessels. The transport of water through the xylem vessels is confirmed by the experiment illustrated in figure 6.9.

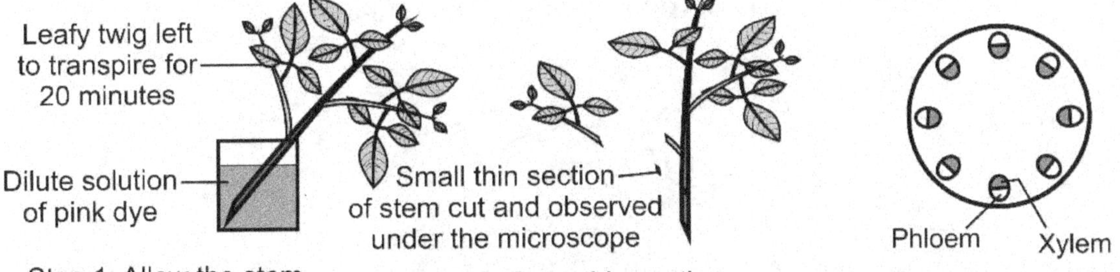

Fig. 6.9: Transport of water through xylem vessels

Cambridge 5090 syllabus specification 6(e) explain the movement of water through the stem in terms of transpiration pull;

Water is transported by xylem vessels from the roots to all parts of the plant. Figure 6.10 describes the movement of water from the roots to the leaves.

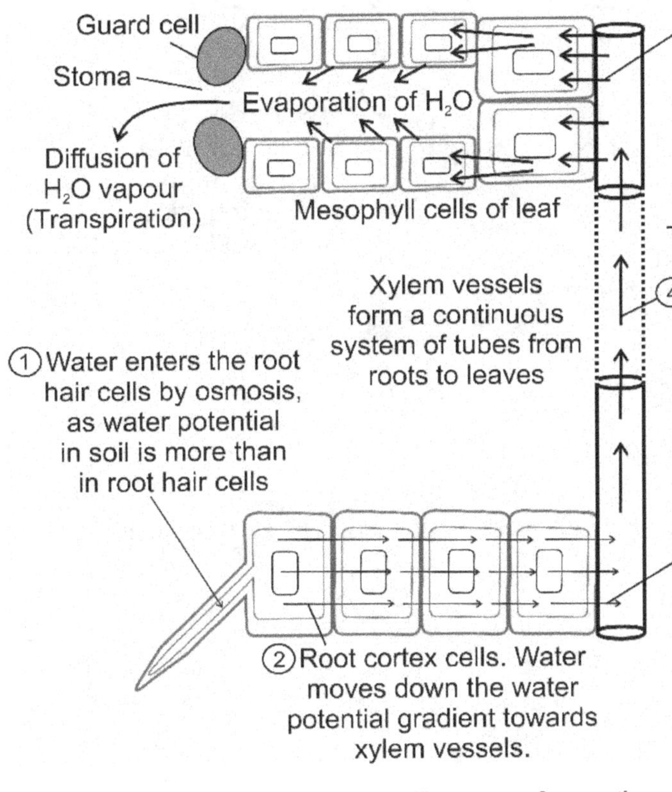

⑤ Mesophyll cells of leaf have lower water potential than xylem vessels of leaf, as mesophyll cells lose water due to transpiration. So, water moves out of the xylem vessels in the leaf by osmosis. This is called the **transpiration pull**.

④ Water moves through the xylem vessels by mass flow or pressure flow. The water flows from high pressure in the roots to low pressure to the leaves. This stream of water is called transpiration stream and is due to cohesion of water molecules.

③ Water potential in root cortex cells is more than in the root xylem. So water enters the root xylem with high pressure, by osmosis. This is called **root pressure** and pushes sap upwards.

Increase in the rate of transpiration, will increase the rate of photosynthesis. This is because more water reaches the leaf cells and is used to increase the rate of photosynthesis.

The open stomata also allow carbon dioxide to enter the leaves for more rapid photosynthesis.

Excess water loss can be prevented by **Wilting**. The guard cells and other leaf cells become flaccid and the stomata close.

Transpiration also cools the plant and prevents it from overheating.

Fig. 6.10: Transport of water through the xylem vessels

The position of xylem vessels and phloem in a dicot stem, root and leaf are shown in figure 6.11.

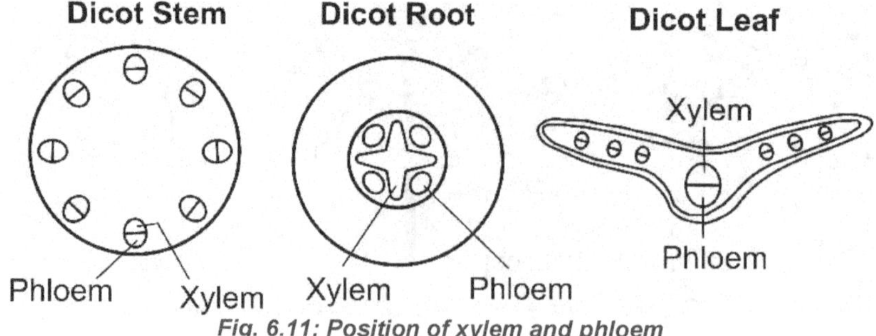

Fig. 6.11: Position of xylem and phloem

Experiment to demonstrate the position of phloem in plants.

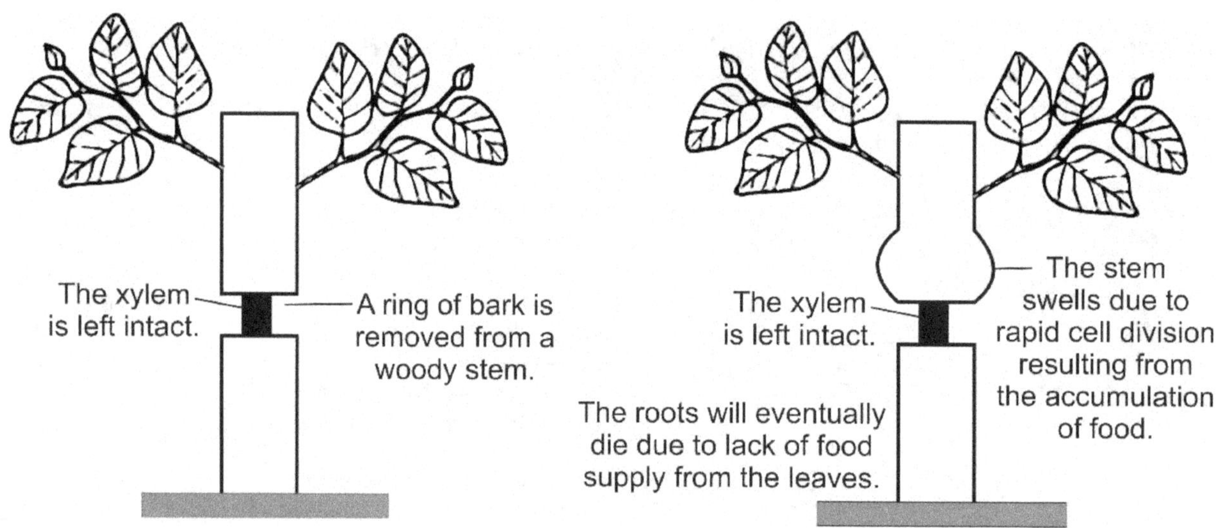

Fig. 6.12: Position of phloem in a woody stem

Function of xylem

The functions of xylem are:

- to transport water and mineral ions from the roots to all other parts of the plant.
- to provide mechanical support to the plant, as the cells contain a very hard waterproof substance called lignin.

Function of phloem

The phloem carries sucrose (sugar) and amino acids to all parts of the plant. The phloem sap also contains ions and water. The sap is carried by mass flow (high pressure to low pressure) from the source to the sink.

Heart: a muscular organ which pumps the blood with high pressure into the arteries. **Arteries:** carry blood at high pressure towards the organs, away from the heart. **Veins:** carry blood at low pressure towards the heart, away from other organs. **Capillaries:** thin walled, permeable vessels which allow exchange of material between blood and cells. **Valves:** prevent backflow of blood and ensure that the blood flows in one direction only. **Blood:** transports food and metabolic wastes. Also helps to fight infections. **Tissue fluid:** a medium of transport between the blood and tissues. **Lymph:** found in lymph vessels and is poured into the veins by lymph vessels.	The circulatory system is made up of the heart, arteries, veins, capillaries, blood, lymph and lymph vessels, as shown in figure 7.1 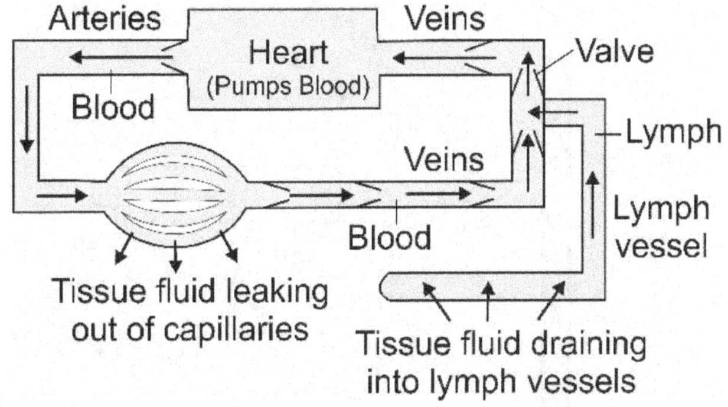 *Fig. 7.1: Components of the circulatory system of humans*

Double circulatory system

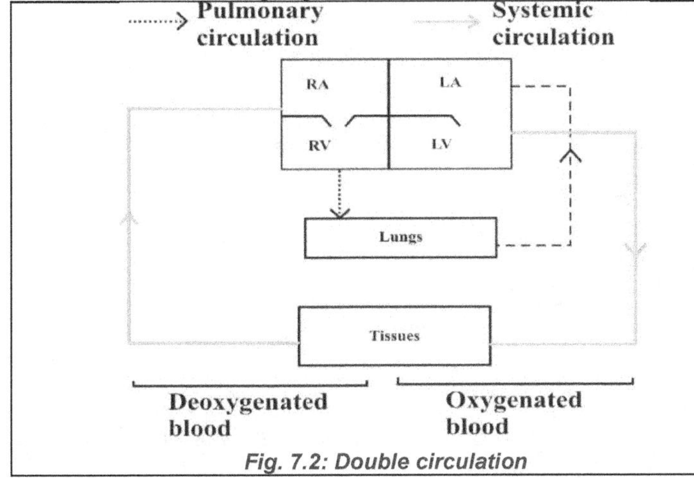 *Fig. 7.2: Double circulation*	The mammalian circulatory system consists of two circulations, as shown in Figure 7.2. • **Systemic circulation** is the pumping of oxygenated blood to all the organ systems of the body. It is a *high pressure circulation* originating from the left ventricle and terminating at the right atrium, which receives deoxygenated blood from the organ systems or tissues. • **Pulmonary circulation** is the pumping of deoxygenated blood to the lungs. It is a *low-pressure circulation* originating from the right ventricle and terminating at the left atrium, which receives oxygenated blood from the lungs.

Double circulation ensures that oxygenated blood does not mix with deoxygenated blood. So, only oxygenated blood is circulated to all the organ systems of the body.

The walls of the **left ventricles are much thicker** as they have to exert a greater pressure to pump blood to all parts of the body, even to the head against the force of gravity. The right ventricles exert a much lower pressure as the lungs are close to the heart and pulmonary tissues are very delicate. High pressure in the pulmonary circulation will cause the lungs to fill with tissue fluid. This will cause the person to drown in their own tissue fluid. The left side of the heart contains only oxygenated blood and the right side contains only deoxygenated blood. The right and left sides are separated by a muscular wall called as the **septum**.

Figure 7.3 shows the major blood vessels associated with the heart, kidneys, lungs and liver.

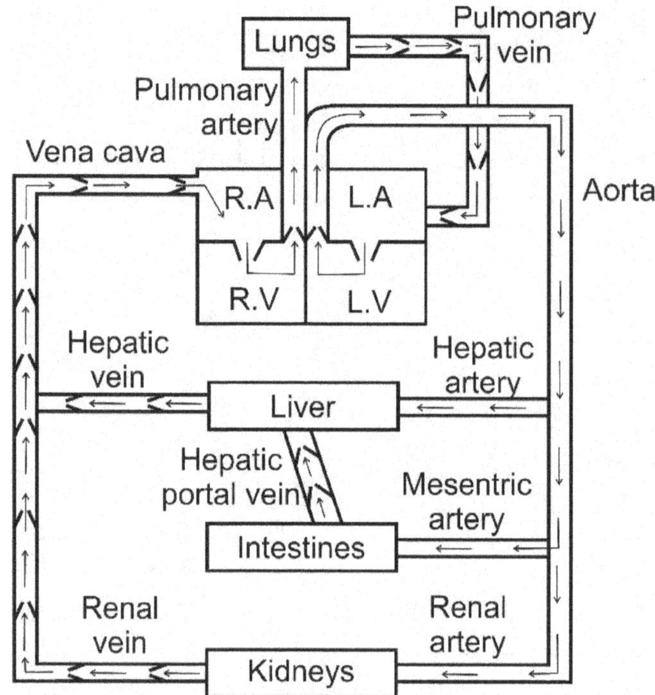

Arteries carry blood at high pressure away from the heart, towards different organs in the body.

Veins carry blood at low pressure towards the heart.

Capillaries connect arteries to veins. They allow the exchange of materials between the blood and the tissues.

Fig. 7.3: Major blood vessels

Structure of mammalian heart

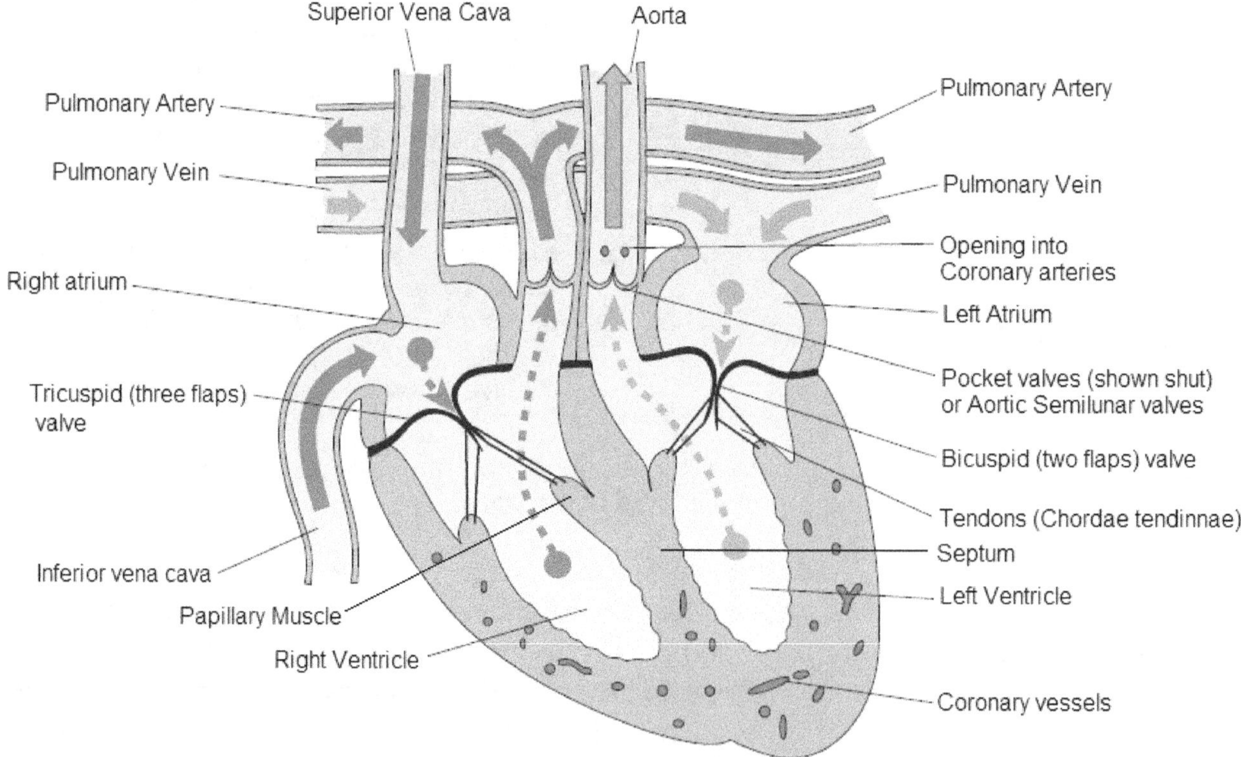

Fig. 7.4: Structure of heart

- The heart is made up of cardiac muscle. It has 4 chambers, 2 upper chambers called atria and 2 lower chambers called ventricles.

- The walls of the **left ventricles are much thicker** as they have to exert a greater pressure to pump blood to all parts of the body, even to the head against the force of gravity.

- The right ventricles exert a much lower pressure as the lungs are close to the heart and pulmonary tissues are very delicate. High pressure in the pulmonary circulation will lead to **pulmonary oedema**, causing the lungs to fill with tissue fluid. This will cause the person to drown in their own tissue fluid.

- The left side of the heart contains only oxygenated blood and the right side contains only deoxygenated blood. The right and left sides are separated by a muscular wall called as the **septum**.

Structure	Function
Superior Vena Cava	Carries deoxygenated blood from the **head** to the **right atrium**.
Inferior Vena Cava	Carries deoxygenated blood from the **body** to the **right atrium**.
Aorta	Carries oxygenated blood from the **left ventricle** to the **body**.
Pulmonary artery	Carries deoxygenated blood from the **right ventricle** to the **lungs**.
Pulmonary vein	Carries oxygenated blood from the **lungs** to the **left atrium**.
Left Atrium	Receives oxygenated blood from pulmonary vein and pumps it into the left ventricle
Right Atrium	Receives deoxygenated blood from vena cava and pumps it into the right ventricle
Left Ventricle	Receives oxygenated blood from the left atrium and pumps it to all the organ systems at high pressure (systemic circulation)
Right Ventricle	Receives deoxygenated blood from the right atrium and pumps it to the lungs at lower pressure (pulmonary circulation)
Semilunar valves	Prevents the backflow of blood from the arteries into the ventricles.
Bicuspid and tricuspid valve (atrio-ventricular valves)	Prevents backflow of blood into the atria during ventricular systole. The closing of these valves also helps to build up a high pressure in the ventricles during ventricular systole, which helps to push blood into the arteries.
Coronary artery	Supplies oxygenated blood to the cardiac muscle
Papillary muscles	Maintains the tension on the *chordae tendinnae*
***Chordae tendinnae* (tendons)**	Prevents the bicuspid and tricuspid valves from flipping over into the atria during ventricular systole.
Septum	Separates the right and left sides of the heart, so that oxygenated blood in the left cannot mix with deoxygenated blood in the right side of the heart.

The Cardiac cycle

The cardiac cycle is the rhythmic contraction and relaxation of the atria and ventricles during one complete heartbeat. The cardiac cycle can be divided in to 3 main stages, as shown in Figures 7.5, 7.6 and 7.7.

1) **Atrial systole, ventricular diastole (0.13 sec)** – The atria contract and push blood into the ventricles through atrio-ventricular valves.	2) **Ventricular systole, atrial diastole (0.3 sec)** – The ventricles contract and pumps blood from ventricles into arteries through semi-lunar valves. Atria are being refilled during this time.	3) **Complete diastole (0.4 sec)** – all four chambers of the heart are relaxed. Bicuspid and tricuspid valves are closed at the start but open later.
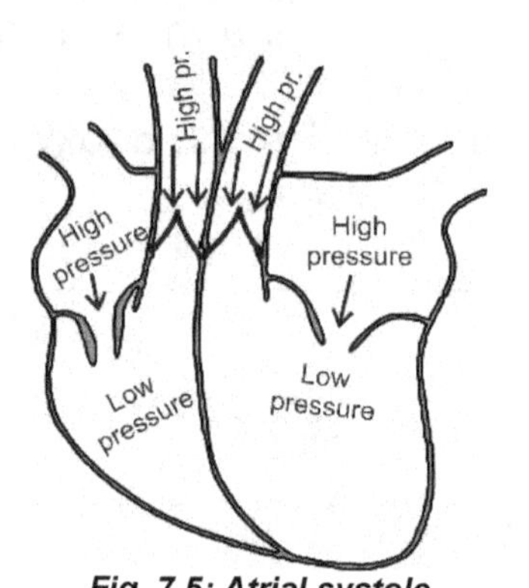Fig. 7.5: Atrial systole	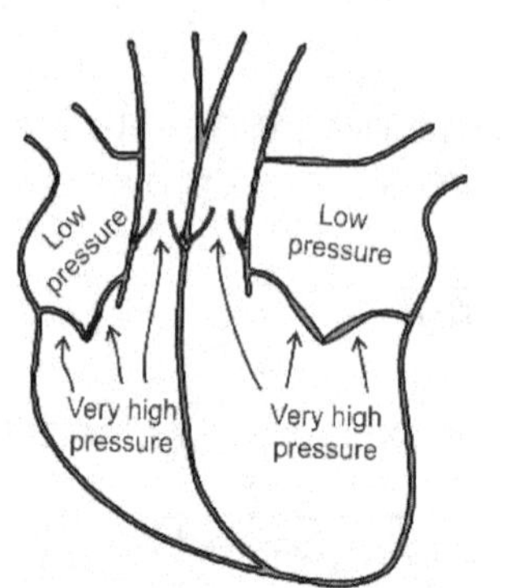Fig. 7.6: Ventricular systole	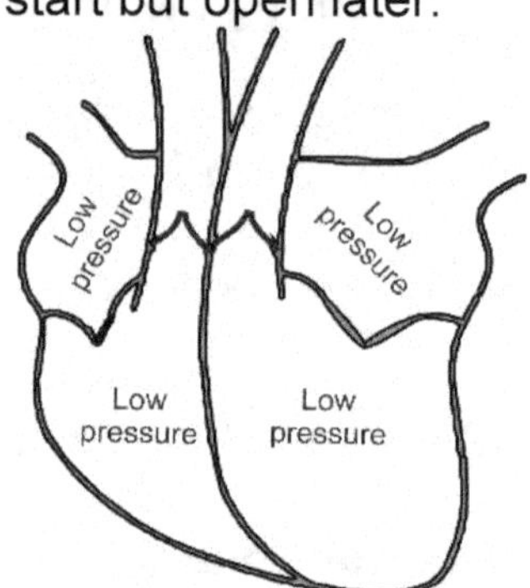Fig. 7.7: Complete diastole

Systole means contraction; **Diastole** means relaxation.

The graph in figure 7.8 shows the pressure changes in the heart of a healthy person at rest.

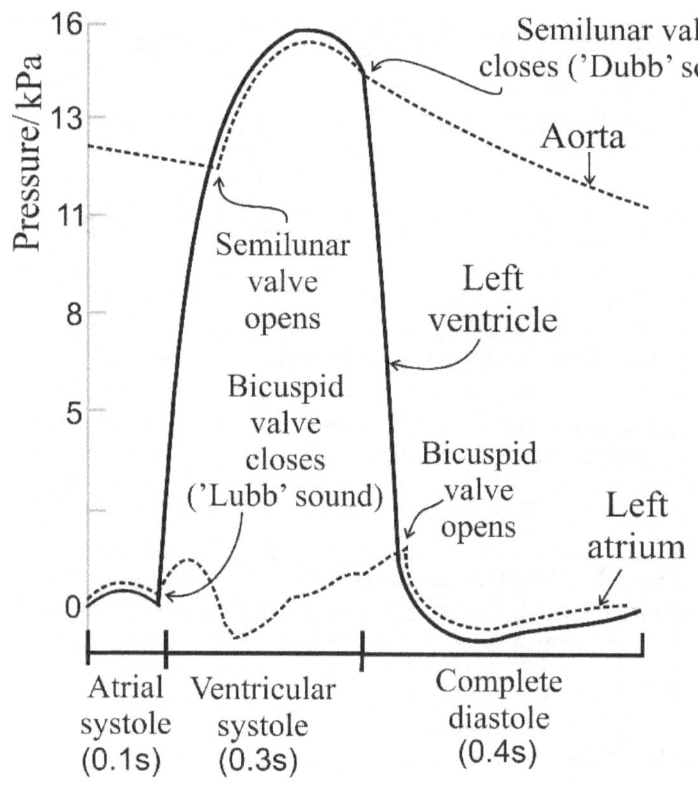

Semilunar valves open when the pressure in the ventricles is greater than in the arteries.

Atrio-ventricular valves open when the pressure in the atria is greater than in the ventricles.

Pressure in the right side of the heart is much lower.

Fig. 7.8: Pressure changes in the heart during the cardiac cycle

Blood vessels

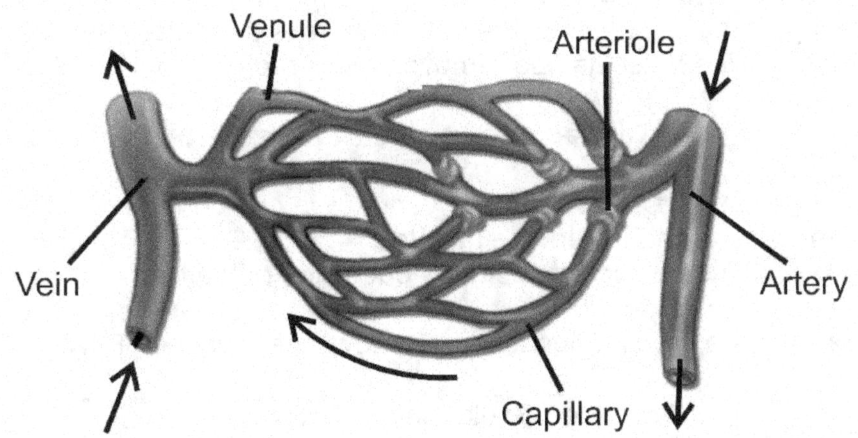

Veins and Venules	Capillaries	Arteries and Arterioles
(Tunica Intima) Endothelium — (Tunica Externa) Collagen and connective tissue — (Tunica Media) Smooth muscle and elastic tissue — Semilunar Valve — Lumen (Blood) — 0.1 to 20 mm	Basement membrane — Endothelium (Tunica Intima) — Red Blood Cell — 8μm	(Tunica Intima) Endothelium — (Tunica Externa) Collagen and connective tissue — (Tunica Media) Smooth muscle and elastic tissue — Lumen (Blood) — 0.1 to 10 mm
Function is to carry blood from tissues to the heart	Function is to allow exchange of materials between the blood and the tissues	Function is to carry blood from the heart to the tissues
Thin walls, mainly collagen, since blood at low pressure	Very thin, permeable walls, only one cell thick to allow easy diffusion of materials	Thick walls with smooth elastic layers to resist high pressure and muscle layer to aid pumping(recoil)
Large lumen to reduce resistance to flow.	Very small lumen. Blood cells must distort to pass through.	Small lumen
Many valves to prevent backflow of blood	No valves	No valves (except in heart)
Blood at low pressure due to large lumen	Blood pressure falls in capillaries as volume of blood decreases due to formation of tissue fluid and the large surface area	Blood at high pressure due to narrow lumen and elastic recoil
Blood usually deoxygenated (except in pulmonary vein and umbilical vein)	Blood changes from oxygenated to deoxygenated (except in lungs)	Blood usually oxygenated (except in pulmonary artery and umbilical artery)

When a healthy individual is at rest, the normal heartbeat rate is about 72 beats per minute. The cardiac output is sufficient enough to supply the muscles with the oxygen they need and to carry away the Carbon dioxide at the required rate. However, as a person starts to exercise there is a greater need for oxygen to be supplied at a greater rate. The carbon dioxide must also be removed more rapidly. This is achieved by an increase in the cardiac output.

Cardiac output = heart rate x stroke volume

Stroke volume is the volume of blood pumped by the ventricle in one contraction.
Heartbeat rate is the number of heartbeats in one minute.
Cardiac output (CO) is the volume of blood pumped by each side of the heart in one minute.

The changes in the heart rate are brought about by the following sequence of events.

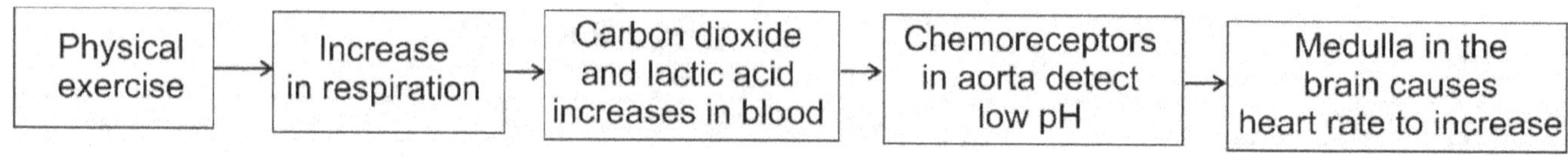

Coronary heart disease (CHD)

Coronary arteries supply oxygen and glucose to the cardiac muscles of the heart. If fatty deposits called atheroma or plaque build-up inside a coronary blood vessel, then the vessel gets blocked and results in lack of blood flow to the cardiac muscles, as shown in figure 7.9. This will lead to lack of oxygen and glucose in the cardiac muscle tissue. The anaerobic conditions will cause the muscles to stop working and damaged cells will die. Lactic acid builds up in the cardiac muscles and results in pain called as **angina pectoris**. This may also lead to a heart attack or myocardial infarction.

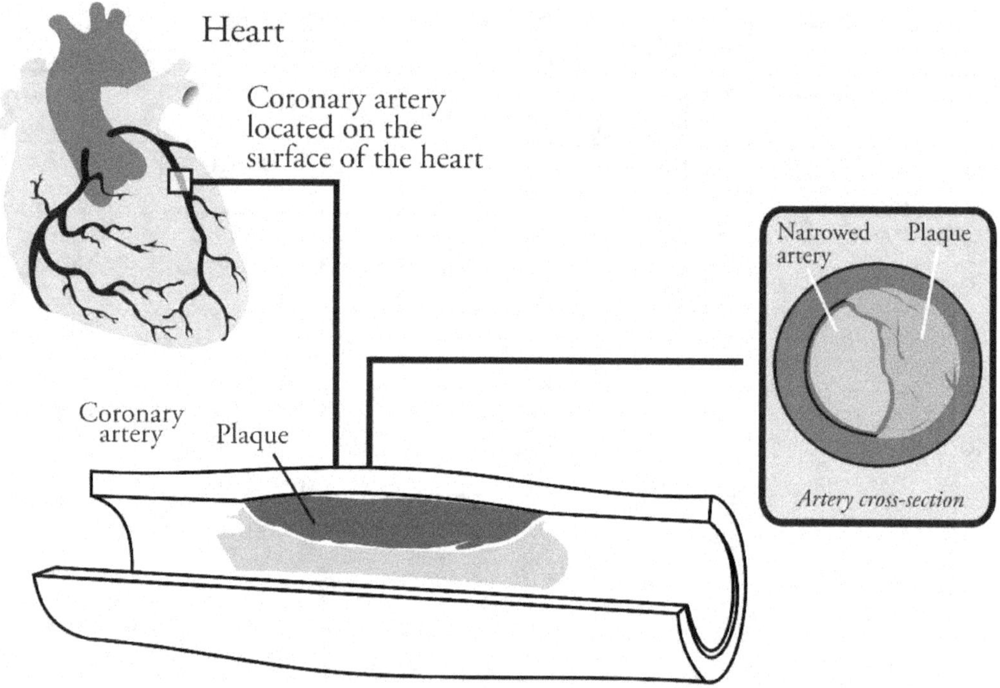

Fig. 7.9: Formation of plaque in coronary arteries

Possible causes of coronary heart diseases

Coronary heart diseases are caused by an improper diet, stress and smoking.

Diet

- Animal fats (saturated fatty acids) increase the risk of plaque formation. This increases the blood pressure and can increase the risk of heart attacks.
- High salt intake can increase blood pressure. Increase in blood pressure can increase the formation of plaque in the coronary arteries and increase the risk of heart attacks.
- Consuming high energy fatty food can lead to obesity. This can increase the risk of plaque formation in the coronary arteries and increase the risk of heart attacks.

Stress

- Stress results in the secretion of adrenaline. Adrenaline increases the blood pressure and heartbeat rate. This increases the risk of atheroma formation in the coronary arteries and increase the risk of heart attacks.

Smoking

- Cigarette smoke contains carbon monoxide. This reduces the oxygen carrying capacity of blood, as the haemoglobin binds irreversibly with carbon monoxide to form carboxyhaemoglobin. This can lead to fatigue of the heart muscles, leading to heart attacks.

Preventive measures

- Consume unsaturated fats (vegetable oils) instead of saturated fats.
- Reduce salt intake.
- Regular exercise and consumption of low energy food can prevent obesity and heart attacks.
- Stop smoking.

> *Cambridge 5090 syllabus specification 7(h) identify red and white blood cells as seen under the light microscope on prepared slides, and in diagrams and photomicrographs;*
> *Cambridge 5090 syllabus specification 7(i) list the components of blood as red blood cells, white blood cells, platelets and plasma;*
> *Cambridge 5090 syllabus specification 7(j) state the functions of blood: • red blood cells – haemoglobin and oxygen transport;*
> *• white blood cells – phagocytosis, antibody formation and tissue rejection; • platelets – fibrinogen to fibrin, causing clotting;*
> *• plasma – transport of blood cells, ions, soluble food substances, hormones, carbon dioxide, urea, vitamins and plasma proteins;*

Figure 7.10 shows the different cells found in blood.

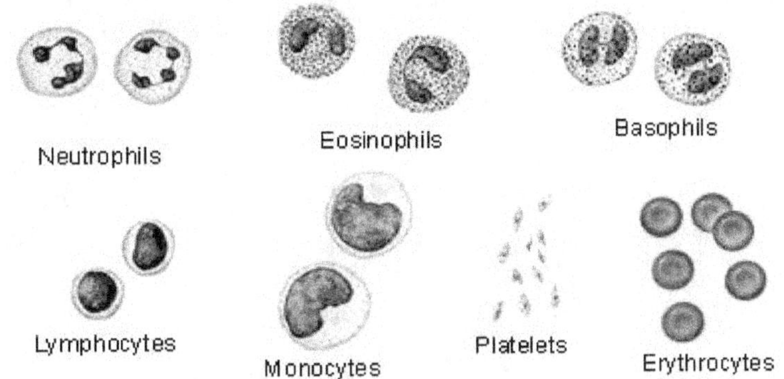

Fig. 7.10: Different types of blood cells and platelets

Component of blood	Function
Red blood cells (Erythrocytes)	Red blood cells contain haemoglobin, which combines with oxygen in the lungs to form oxy-haemoglobin. The oxy-haemoglobin dissociates in the tissues to release oxygen for tissues to respire.
White blood cells (Lymphocytes)	The lymphocytes produce antibodies which help to identify and destroy pathogens in the body. It also results in tissue rejection of foreign tissues.
White blood cells (Phagocytes)	Phagocytes engulf pathogens in the body and destroy the pathogens.
Platelets	Platelets are fragments of cells. They play a role in blood clotting.
Plasma	Plasma is the liquid component of blood. It transports blood cells, ions, soluble food substances, hormones, carbon dioxide, urea, vitamins and plasma proteins throughout the body.

Blood clotting

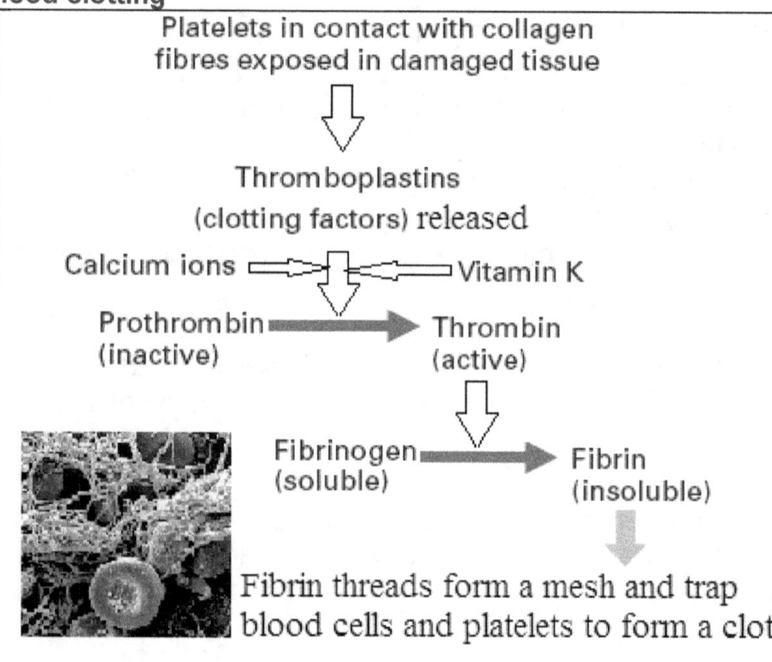

Platelets stick to a site of damaged tissue. Thromboplastin is released from this area, which results in a plasma protein called prothrombin being converted to thrombin. Thrombin acts as an enzyme and catalyses the conversion of the soluble plasma protein fibrinogen into long, insoluble strands of the protein fibrin These strands form a mesh that traps red blood cells to form a blood clot. Blood clotting minimizes blood loss following injury. The blood coagulates to form a solid plug (clot) made of cells trapped in a fibrous network. The **clot**

- **prevents further blood loss,**
- **reduces the risk of pathogens (harmful micro-organisms) entering the body, and**
- **provides a framework for the repair of damaged tissue.**

Cambridge 5090 syllabus specification 7(k) describe the transfer of materials between capillaries and tissue fluid.

FORMATION AND REABSORPTION OF TISSUE FLUID

Tissue fluid is a transport medium between blood in the capillaries and tissues. It carries nutrients from capillaries to tissues and other wastes flow back from tissue to capillaries. These substances can pass from capillaries to tissue fluid and vice versa by ultrafiltration, osmosis and diffusion. Tissue fluid is formed from blood at the arterial end of the capillaries. The high blood pressure, thin and porous walls of the capillaries causes the liquid component of blood (except large proteins) to be squeezed out of the capillaries. About 85% of the tissue fluid is reabsorbed at the venous ends of the capillaries mainly by osmosis. Solutes move along the Pressure gradient caused by difference in water potential between the inside and outside of the capillaries. The remaining 15% of tissue fluid is collected by lymph capillaries which pour lymph back into the veins.

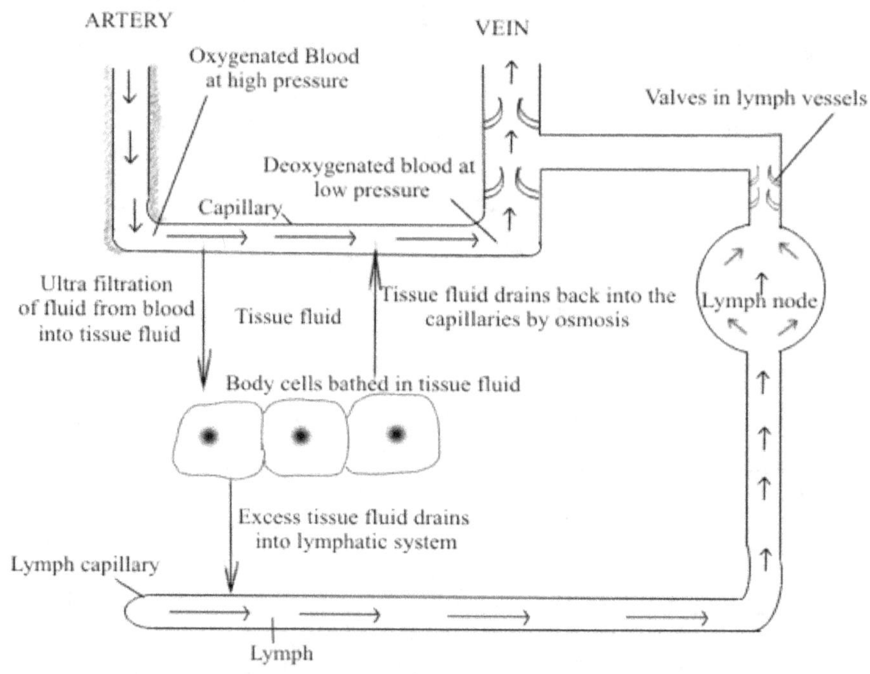

Fig. 7.11: Tissue fluid and lymph

> *Cambridge 5090 syllabus specification 8(a) define respiration as the release of energy from food substances in all living cells;*
> *Cambridge 5090 syllabus specification 8(b) define aerobic respiration as the release of a relatively large amount of energy by the breakdown of food substances in the presence of oxygen;*
> *Cambridge 5090 syllabus specification 8(c) state the equation (in words or symbols) for aerobic respiration;*
> *Cambridge 5090 syllabus specification 8(d) state the uses of energy in the human body: muscle contraction, protein synthesis, cell division, active transport, growth, the passage of nerve impulses and the maintenance of a constant body temperature;*

Respiration is a **chemical process** that occurs in all living cells. Enzymes within the cells are used to breakdown glucose and other substrates to release energy that can then be used for metabolic reactions in the body.

Aerobic respiration

Aerobic respiration occurs within living cells **when oxygen is available** in abundance. It **produces a large amount of energy** in the form of an energy rich molecule called ATP (Adenosine Triphosphate) and **heat energy**. The ATP is used to provide energy for all other metabolic reactions in the cells. The equation below shows the products of aerobic respiration.

$$\text{Glucose} + \text{Oxygen} \longrightarrow \text{Carbon dioxide} + \text{Energy (38 ATP)}$$

$$C_6H_{12}O_6 + 6O_2 \longrightarrow 6CO_2 + 6H_2O + \text{Energy (38 ATP)}$$

Uses of energy released from respiration

- ATP is used for muscle contraction.
- ATP is used to make proteins within the cells of the body.
- ATP is used for cell division and growth.
- ATP is used for active transport.
- ATP is used for the conduction of nerve impulses.
- Respiration also releases thermal energy which is used in the maintenance of a constant body temperature;

> *Cambridge 5090 syllabus specification 8(e) define anaerobic respiration as the release of a relatively small amount of energy by the breakdown of food substances in the absence of oxygen;*
> *Cambridge 5090 syllabus specification 8(f) state the equation (in words or symbols) for anaerobic respiration in humans and in yeast;*

Anaerobic respiration

Anaerobic respiration occurs within living cells **in the absence of oxygen**. It produces a small amount of energy in the form of an energy rich molecule called ATP (Adenosine Triphosphate). The ATP is used to provide energy for all other metabolic reactions in the cells.

The equation below shows the products of anaerobic respiration **in Yeast cells**. This is also called fermentation and is used for the production of ethanol and bread.

$$\text{Glucose} \longrightarrow \text{Carbon dioxide} + \text{Ethanol} + \text{Energy (2 ATP)}$$

$$C_6H_{12}O_6 \longrightarrow 2CO_2 + 2C_2H_5OH + \text{Energy (2 ATP)}$$

The equation below shows the products of anaerobic respiration **in muscle cells**.

$$\text{Glucose} \longrightarrow \text{Lactic acid} + \text{Energy (2 ATP)}$$

$$C_6H_{12}O_6 \longrightarrow 2C_3H_6O_3 + \text{Energy (2 ATP)}$$

Effects of lactic acid production in muscles

- During vigorous exercise, the rate of oxygen supply to the muscle cells may be less than the demand. This leads to an oxygen debt and the muscle cells start respiring anaerobically.
- Lactic acid builds up in the muscle cells.
- pH of muscle cells is lowered due to the lactic acid.
- The low pH denatures proteins in the muscle cells and results in severe pain and cramping of muscles.

The table below shows the composition of gases in inhaled and exhaled air.

Gas	Percentage in inhaled air (Approximate values)	Percentage in exhaled air (Approximate values)
Oxygen	21	16
Carbon dioxide	0.04	4
Nitrogen	78	78
Water vapour	Variable	Saturated

Investigating the difference in concentration of carbon dioxide between exhaled and inhaled air

Set up two tubes A and B as shown in figure 8.1. The lime water at the start of the experiment is colourless. As the persons inhales and exhales into and out of tube A and B respectively, the limewater in tube B turns milky white much faster than in tube A. this proves that exhaled air has a higher concentration of carbon dioxide than inhaled air.

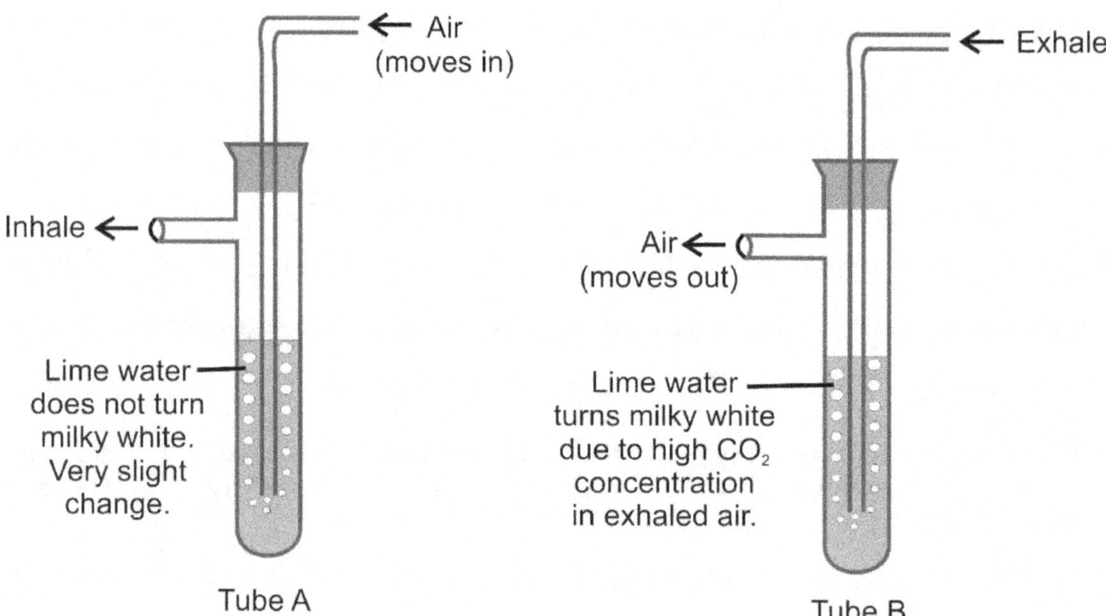

Fig. 8.1: Investigating Carbon dioxide concentration in inhaled and exhaled air

Lung volumes can be measured by using a spirometer, shown in figure 8.2. It is not necessary to learn the details of the working, as the specification only requires you to interpret spirometer traces before and after exercise.

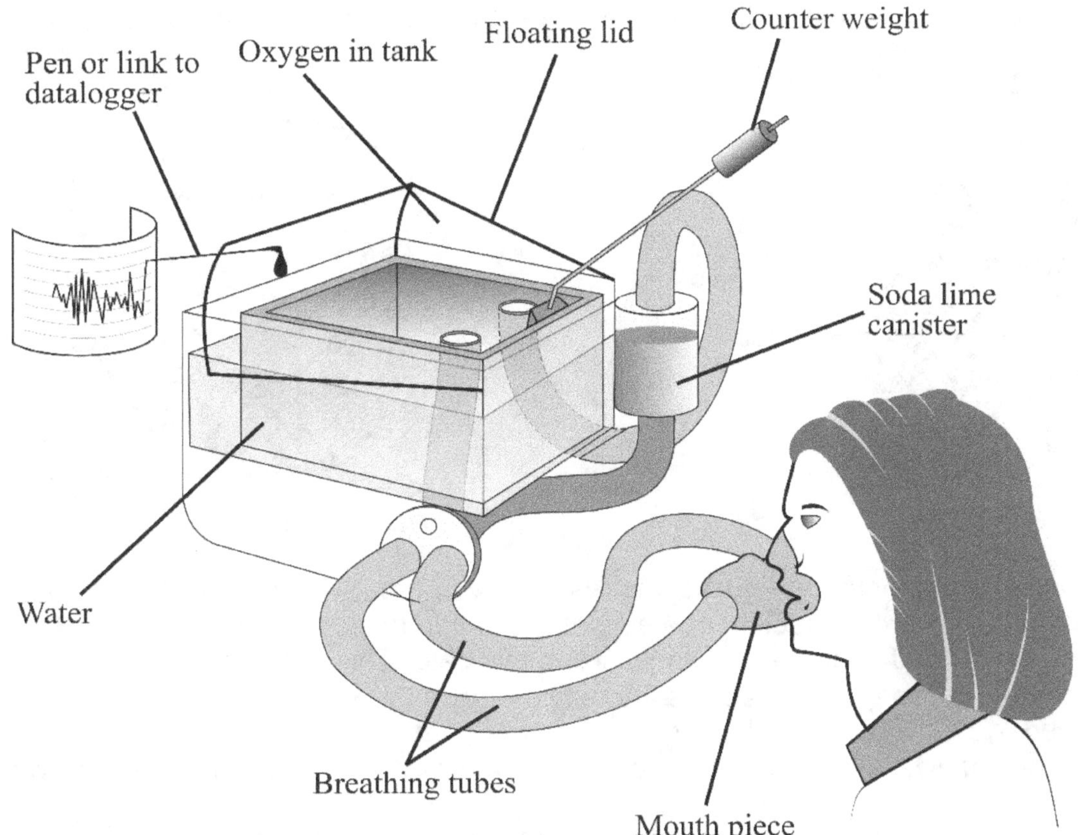

Fig. 8.2: Spirometer

The spirometer produces a graph which can be used to interpret the breathing rate and depth.

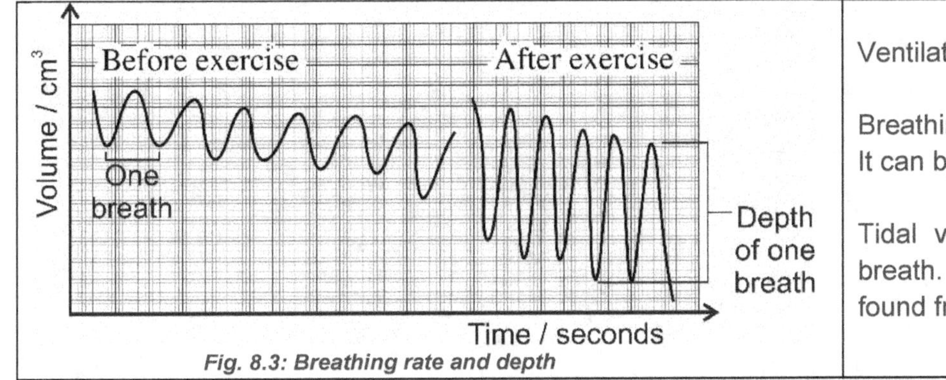

Fig. 8.3: Breathing rate and depth

Ventilation rate = Breathing rate x Tidal volume

Breathing rate is the number of breaths per minute. It can be found from the spirometer trace.

Tidal volume is the volume of air inhaled per breath. It is the depth of breathing. It can also be found from the spirometer trace in figure 8.3.

The spirometer traces are measured before exercise (at rest) and again after a period of exercise. The traces show that both breathing rate and depth increase during exercise. The reasons for the changes are explained in figure 8.4.

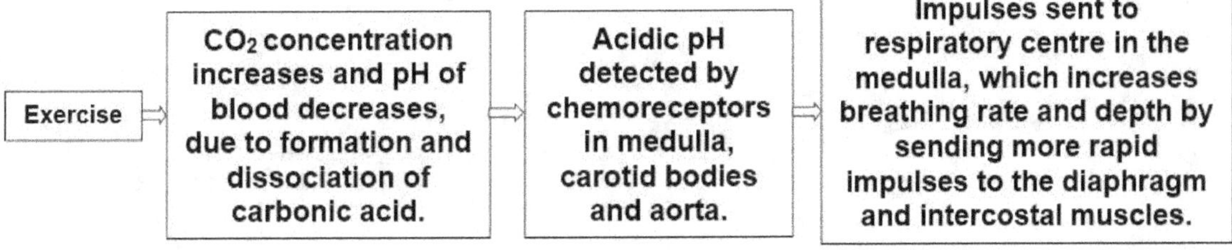

Fig. 8.4: Regulation of breathing rate and depth

Cambridge 5090 syllabus specification 8(j) identify on diagrams and name the larynx, trachea, bronchi, bronchioles, alveoli and associated capillaries;
Cambridge 5090 syllabus specification 8(k) state the characteristics of, and describe the role of, the exchange surface of the alveoli in gas exchange;

Figure 8.5 and 8.6 show the various structures in the human lung and the capillaries surrounding the alveoli.

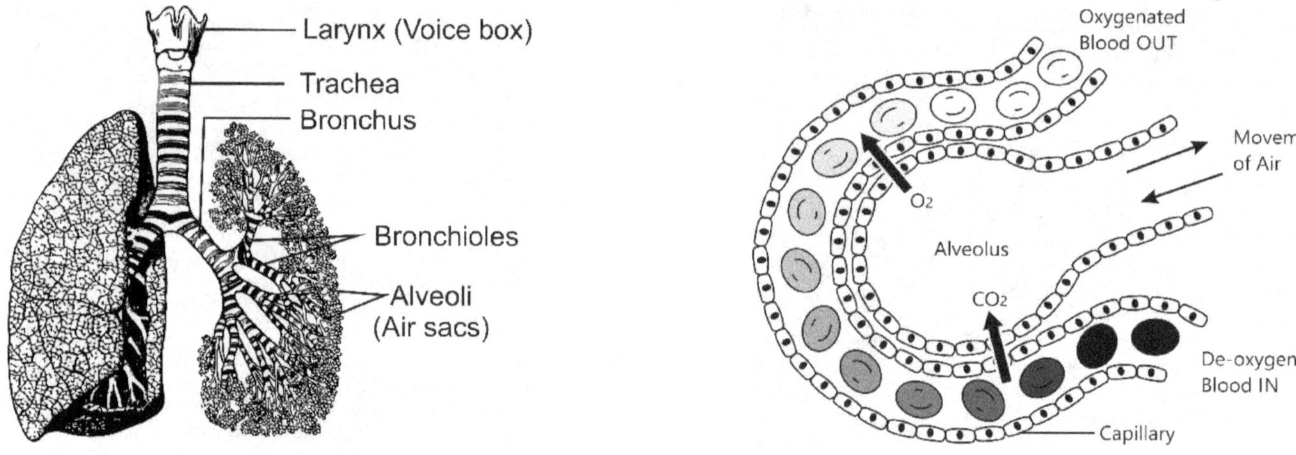

Fig. 8.5: Structure of lungs

Fig. 8.6: Structure of alveoli and blood capillaries

In humans the gas exchange organ system is the respiratory or breathing system. The main features are:

- The walls of the alveoli are composed of a single layer of flattened epithelial cells, as are the walls of the capillaries, so gases need to diffuse through just two thin cells. This decreases the **diffusion distance**.
- The **steep concentration gradient** across the respiratory surface is maintained in two ways: by blood flow on one side and by air flow on the other side.
- There are millions of alveoli and blood capillaries, which provide **an extremely large surface** for gases to be exchanged.
- The flow of air in and out of the alveoli is called **ventilation** and has two stages: **inspiration** (or inhalation) and **expiration** (or exhalation). Lungs are not muscular and cannot ventilate themselves, but instead the whole **thorax** moves and changes size, due to the action of two sets of muscles: the **intercostal muscles** and the **diaphragm**.

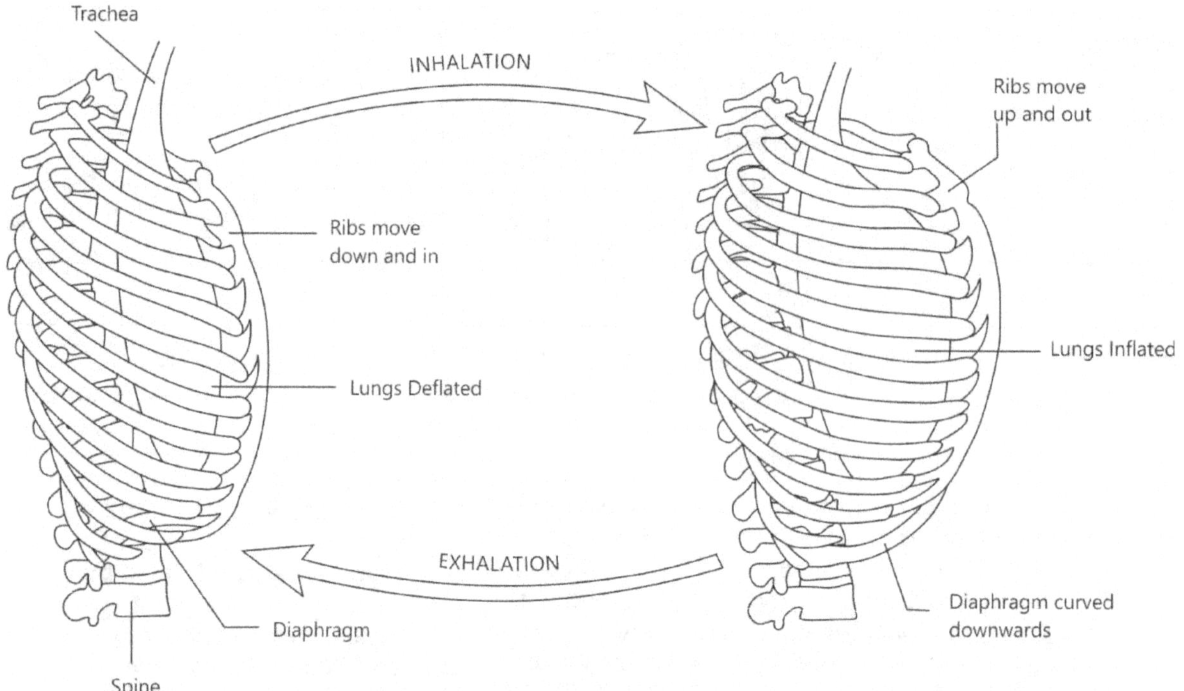

Fig. 8.7: Ribs, sternum and diaphragm assist in breathing

Cambridge 5090 syllabus specification 8(I) describe the role of cilia, diaphragm, ribs and intercostal muscles (external and internal) in breathing.

Role of cilia in breathing

The cilia are found on epithelial cells lining the trachea. The Goblet cells produce mucus. The sticky mucus traps bacteria, viruses and dust particles from the inhaled air. The cilia then sweep the mucus and trapped particles towards the pharynx, as shown in figure 8.8. The mucus can be swallowed. The pathogens are destroyed by hydrochloric acid in the stomach. This helps to reduce the risk of lung infections.

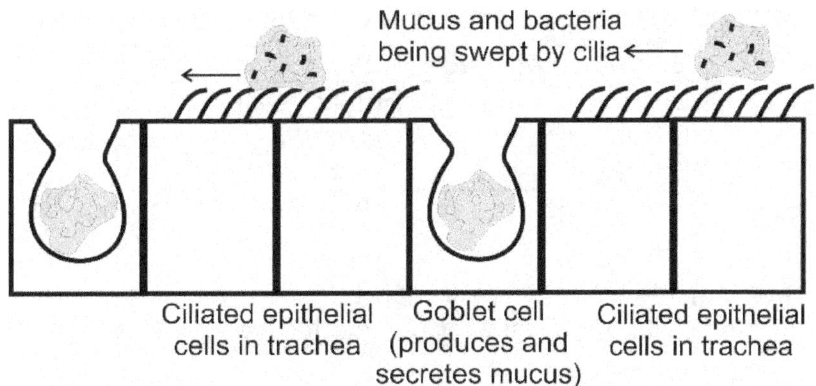

Fig. 8.8: Role of cilia in the trachea

Role of intercostal muscles and diaphragm in breathing

The contraction of intercostal muscles and diaphragm are used to change the volume of the thoracic cavity during breathing. This creates a pressure gradient and results in exhalation or inhalation as described in the table at the end of the page.

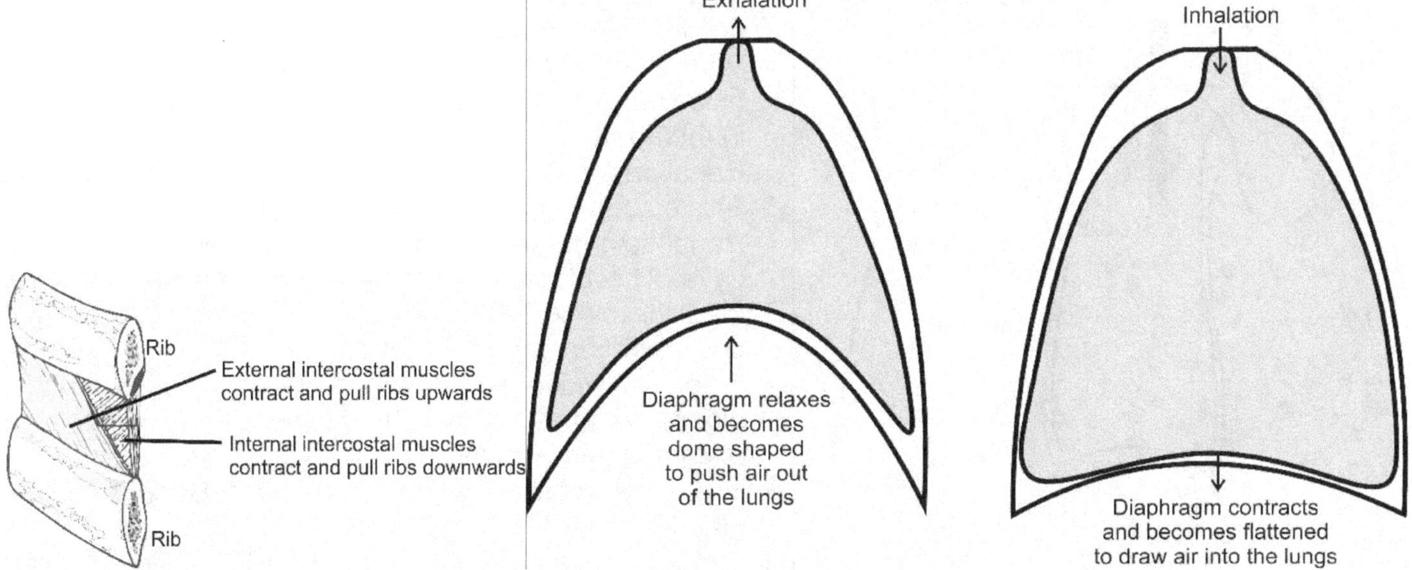

Fig. 8.9: Role of diaphragm and intercostal muscles

During inspiration / inhalation	During exhalation / expiration
• External intercostal muscles contract.	• Internal intercostal muscles contract.
• Internal intercostal muscles relax.	• External intercostal muscles relax.
• Ribs and sternum pulled outwards/upwards.	• Ribs and sternum pulled inwards.
• Diaphragm contracts, becomes flattened.	• Diaphragm becomes dome shaped.
• Volume of thoracic cavity increases causing pressure to become less than atmospheric pressure, so that air rushes into lungs.	• Volume of thoracic cavity decreases.
	• Pressure in lungs becomes greater than atmospheric pressure, so air is pushed out of the lungs.

Excretion is the removal of toxic materials and waste products of metabolism from the bodies of organisms. Some of the excretory products and the organs of excretion are listed below.

Excretory product	Excretory organ
Urea	Kidneys, skin
Carbon dioxide	Lungs
Excess water	Kidneys, skin

The removal of carbon dioxide from the lungs is an example of excretion, as carbon dioxide is a waste product of respiration.

The removal of faeces is not considered to be excretion because the faeces are not products of metabolic reactions of the body. The faeces simply consist of undigested cellulose or lignin which is consumed as a constituent of food.

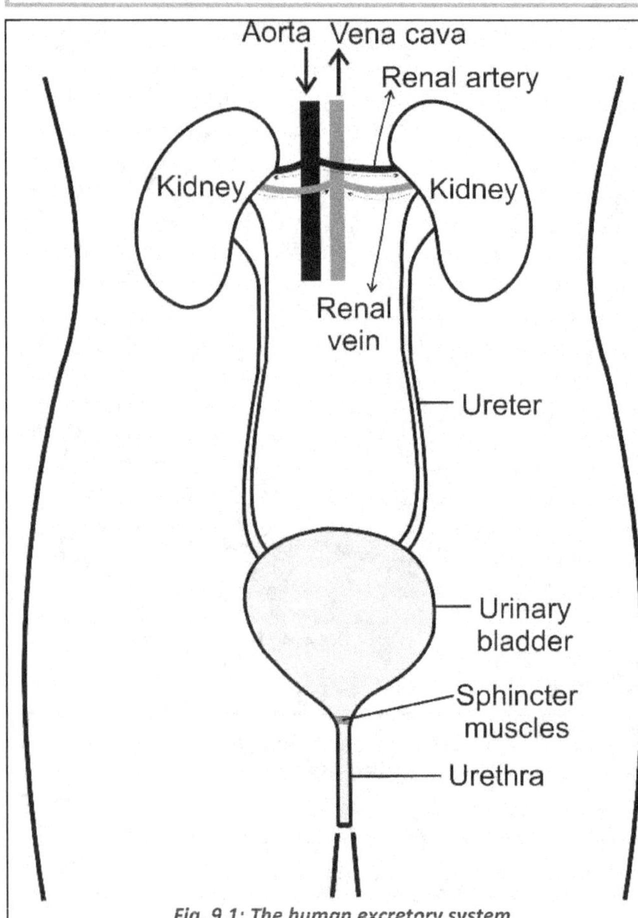

Fig. 9.1: The human excretory system

Functions of the excretory system in humans

- Aorta brings blood towards the kidneys. This blood is rich in urea and oxygen.
- Vena cava carries blood away from the kidneys. The blood has less urea, oxygen and glucose than the blood in the aorta.
- Renal artery carries blood from the aorta to the kidneys. This blood is rich in urea and oxygen.
- Renal vein carries blood from the kidneys to the vena cava. The blood has less urea, oxygen and glucose than the blood in the renal artery. It also has a high concentration of carbon dioxide.
- Kidneys remove excess water, urea and salts from the blood, to form urine. The kidney is an organ for excretion and osmoregulation.
- The ureter carries the urine from the kidneys to the bladder by peristalsis.
- The bladder stores urine temporarily. It is a muscular organ which can contract and push the urine out during urination.
- The sphincter muscles control the release of urine from the bladder. It is under voluntary control in adults.
- Urethra is a tube which carries the urine from the bladder to the outside. In males, it is a common passage for urine and semen.

Dialysis in kidney machines

Dialysis is carried out in people who suffer from kidney failure. During dialysis by using kidney machines, the urea, excess water and salts diffuse through a membrane and leave the blood. Large molecules (e.g. protein) remain in the blood, as shown in figure 9.2.

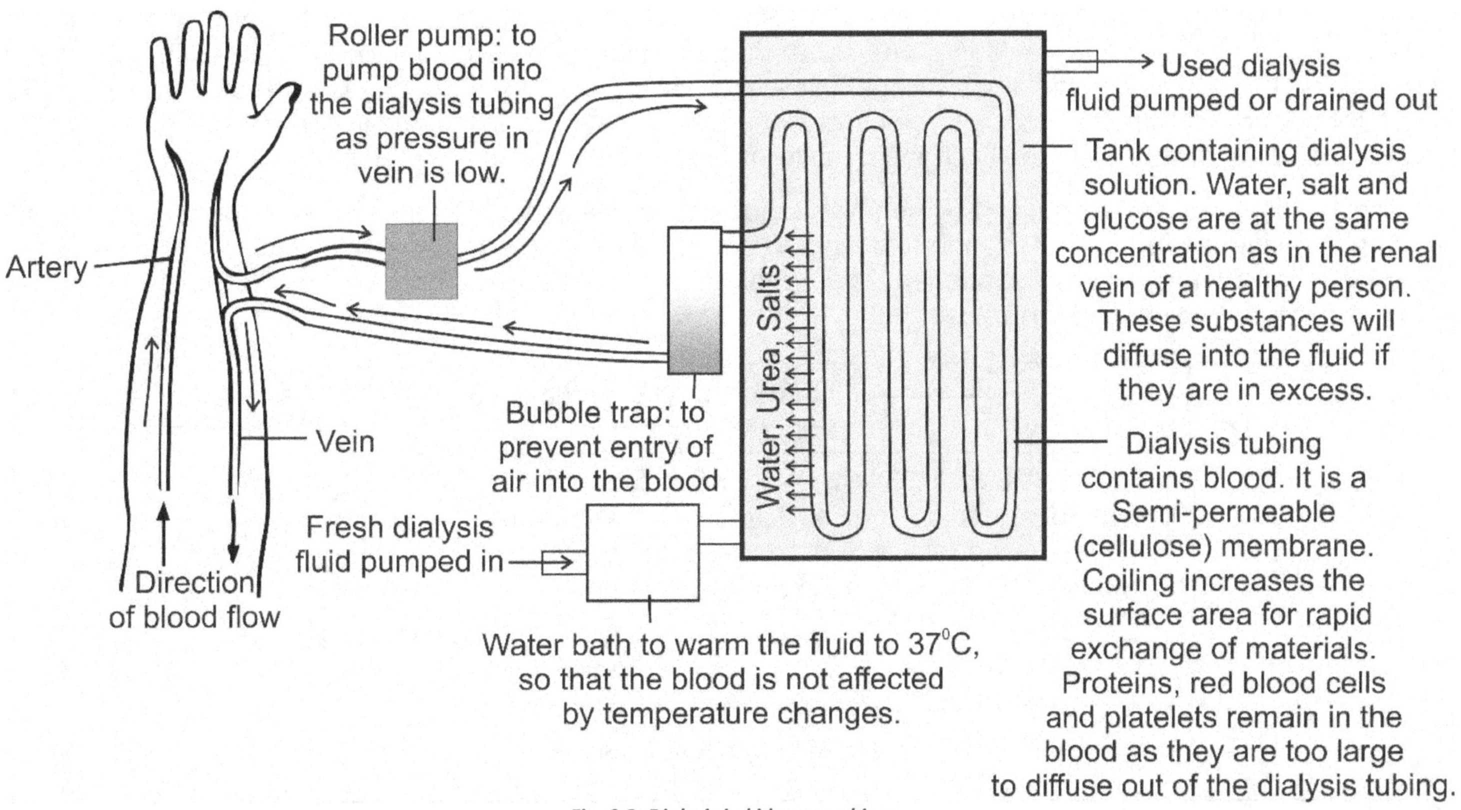

Fig. 9.2: Dialysis in kidney machines

Alternatively, peritoneal dialysis can be carried out by using the peritoneal membrane in the abdominal cavity as a semipermeable membrane. Urea, excess water and salts diffuse through the peritoneal membrane and leave the blood. Large molecules (e.g. protein) remain in the blood, as shown in figure 9.3.

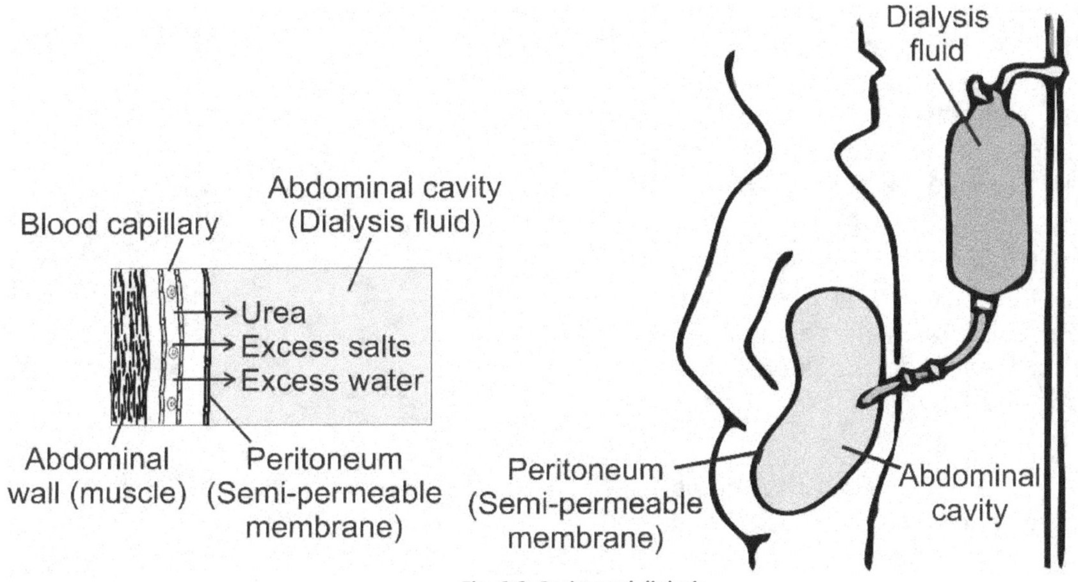

Fig. 9.3: Peritoneal dialysis

Chapter Ten
Homeostasis

Cambridge 5090 syllabus specification 10(a) define homeostasis as the maintenance of a constant internal environment;
Cambridge 5090 syllabus specification 10(b) explain the concept of control by negative feedback;
Cambridge 5090 syllabus specification 10(c) identify, on a diagram of the skin, hairs, sweat glands, temperature receptors, blood vessels and fatty tissue;

Homeostasis (a Greek term meaning same state), is the maintenance of constant conditions in the internal environment of the body despite large swings in the external environment. Functions such as blood pressure, body temperature, respiration rate, and blood glucose levels are maintained within a range of normal values around a set point despite constantly changing external conditions.

Homeostatic systems are maintained by negative feedback mechanisms, sometimes called negative feedback loops. In negative feedback, any change or deviation from the normal range of function is opposed, or resisted. The change or deviation in the controlled value initiates responses that bring the function of the organ or structure **back to within the normal range.**

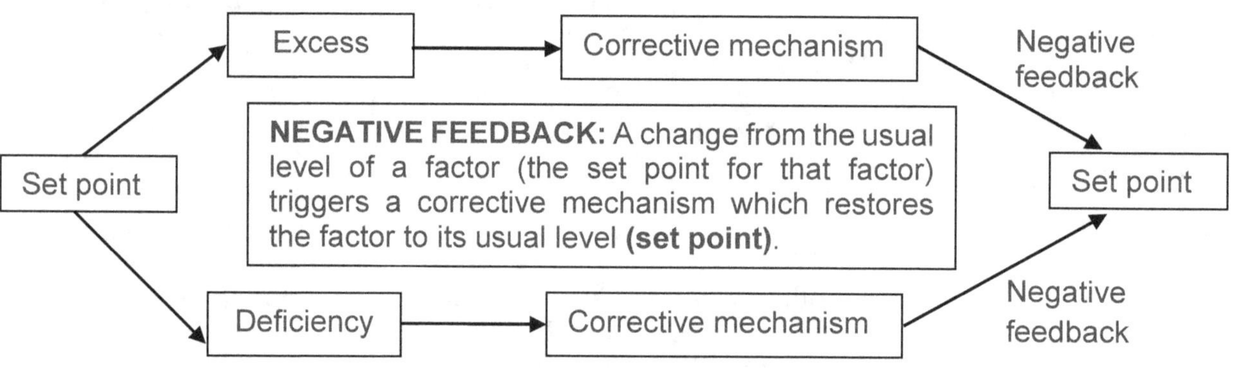

Fig. 10.1: Negative feedback mechanism

Structure of skin

The skin has three layers:

1. Epidermis

The epidermis is the relatively thin, tough, outer layer of the skin. It consists of keratinised cells which are gradually shed and are replaced by newer cells pushed up from below. It is relatively waterproof and, when undamaged, prevents most bacteria, viruses, and other foreign substances from entering the body. The epidermis contains melanocytes, which produce the melanin to give the skin its colour.

2. Dermis

The dermis is a thick layer of fibrous and elastic tissue that gives the skin its flexibility and strength. The dermis contains nerve endings, sweat glands and oil (sebaceous) glands, hair follicles, and blood vessels.

3. Fat layer

Below the dermis lies a layer of fat that helps insulate the body from heat and cold, provides protective padding, and serves as an energy storage area.

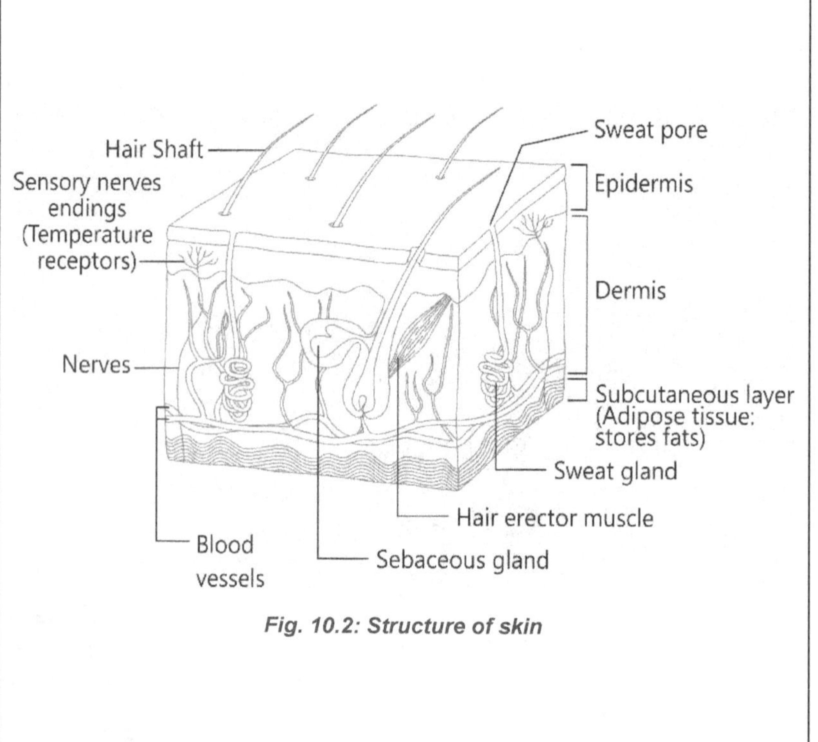

Fig. 10.2: Structure of skin

Negative feedback and thermoregulation

The process by which the body temperature is regulated is illustrated in the following flow chart.

Normal body temperature (37°C)

Decrease in temperature due to
- Physical inactivity
- Lack of clothing
- Heat loss to cold surrounding

↓

Decrease in temperature is detected by temperature **receptors in the hypothalamus and skin**

↓

Impulses sent to **temperature control centre in the hypothalamus (Thermo-regulatory centre)**

↓

Thermo-regulatory centre brings about the following changes:
Vasoconstriction - smooth muscles of arterioles contract and less blood flows into peripheral capillaries. Heat loss by radiation decreases.

Hair erector muscles contract and pull the hair follicles. The hair becomes erect and traps air around the skin. Air is a good insulator and reduces heat loss from the body.

Decreased sweat production results in less heat loss from the skin.

Increased metabolism in the liver and shivering produce more heat.

Increase in temperature due to
- High physical activity
- Warm clothing
- Heat gained from warm surrounding

↓

Increase in temperature is detected by temperature **receptors in the hypothalamus and skin**

↓

Impulses sent to **temperature control centre in the hypothalamus (Thermo-regulatory centre)**

↓

Thermo-regulatory centre brings about the following changes:
Vasodilation - smooth muscles of arterioles relax and more blood flows into peripheral capillaries. Heat loss by radiation increases.

Hair erector muscles relax. The hair lies flat on the skin and does not trap air around the skin. Heat loss from the body increases.

Increased sweat production results in more heat loss from the skin.

Decreased metabolism in the liver and no shivering. Less heat produced.

Normal body temperature (37°C)

The nervous system helps us to respond to various stimuli. The nervous system consists of:
- The central nervous system (the brain and spinal cord), and
- The peripheral nervous system (12 pairs of cranial nerves and 31 pairs of spinal nerves)

The central nervous system receives nerve impulses from the sensory neurones and sends impulses to various organs by the motor neurones.

Structure of the human brain

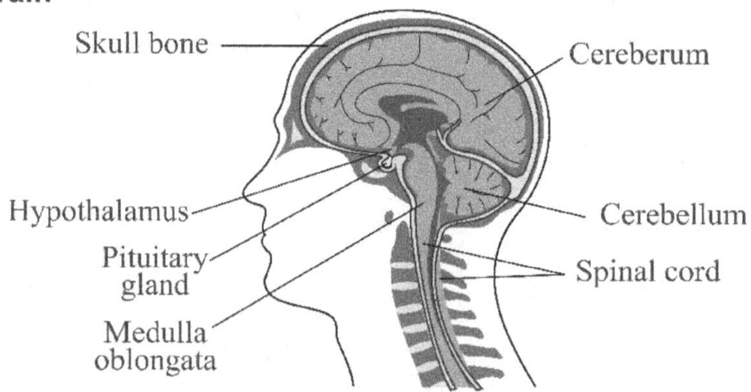

Fig. 11.1: Structure of brain

Functions

- ✓ **Cerebrum**: receives impulses, controls contraction of voluntary muscles, controls mental activities like speech, memory, emotions and other conscious activities.

- ✓ **Cerebellum**: maintenance of balance, posture of body and coordination of skeletal muscles.

- ✓ **Medulla oblongata**: controls rate of heartbeat, breathing rate, reflexes like coughing, sneezing and vomiting.

- ✓ **Hypothalamus**: plays an essential role in osmoregulation, maintaining body temperature and synthesis of hormones of posterior pituitary gland.

- ✓ **Pituitary gland**: it is called the master gland, as it secretes hormones that control the release of hormones from other endocrine glands. For example, Thyroid stimulating hormone from the pituitary gland controls the secretion of thyroid from the thyroid glands.

- ✓ **Spinal cord**: It is also a part of the central nervous system. There are 31 pairs of spinal nerves which connect the spinal cord to various organs in the body. The spinal cord is associated with controlling involuntary reflex actions.

- ✓ **Nerves**: There are 12 pairs of cranial nerves, which connect the brain to various organs. The 31 pairs of spinal nerves connect the spinal cord to various organs as well. The nerves are made up of neurones and carry impulses between the central nervous system (brain and spinal cord) and other organs.

Figure 11.2 and 11.3 shows the structures found in the human eye.

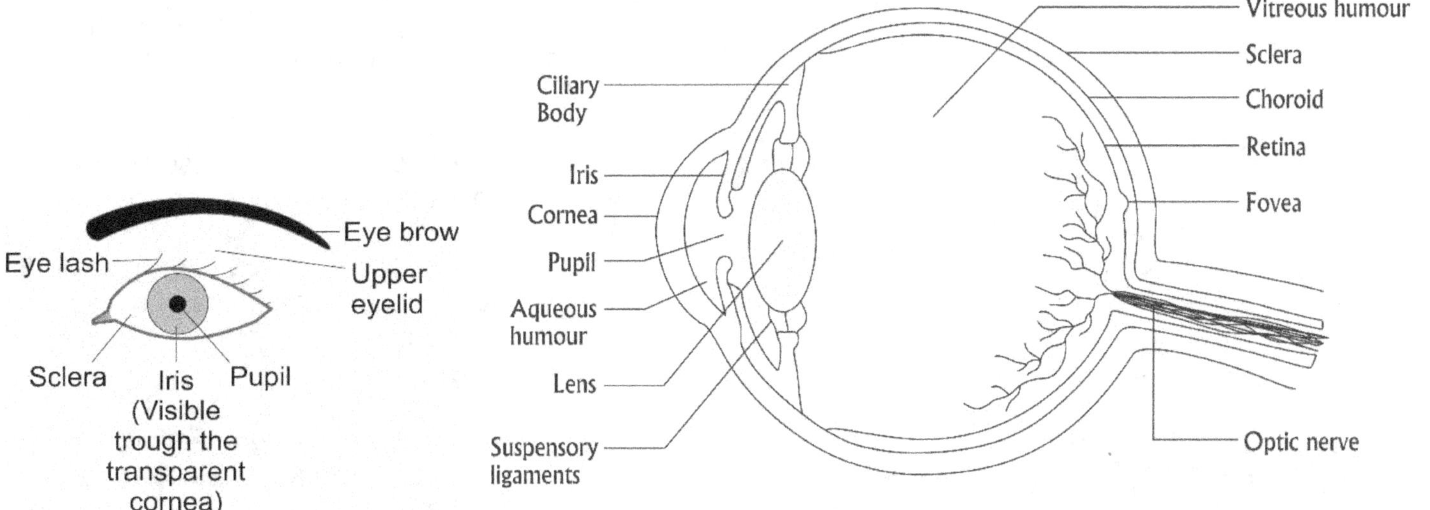

Fig. 11.2: External structure of eye   *Fig. 11.3: Cross sectional structure of the eye*

The main parts of the eye and their function

Part	Description	Function
Cornea	The transparent convex structure covering the iris.	refracts light as it enters the eye
Iris	The coloured region of the eyes. Its muscles contract and relax to alter the size of the pupil.	controls the amount of light entering the pupil
Lens	Transparent, bi-convex, flexible disc behind the iris attached by the suspensory ligaments to the ciliary muscles.	focuses light onto the retina
Retina	The lining of the back of eye containing rods and cones.	contains the light receptors
Optic nerve	Bundle of sensory neurones at back of eye.	carries impulses from the eye to the brain

Pupil dilation and contraction

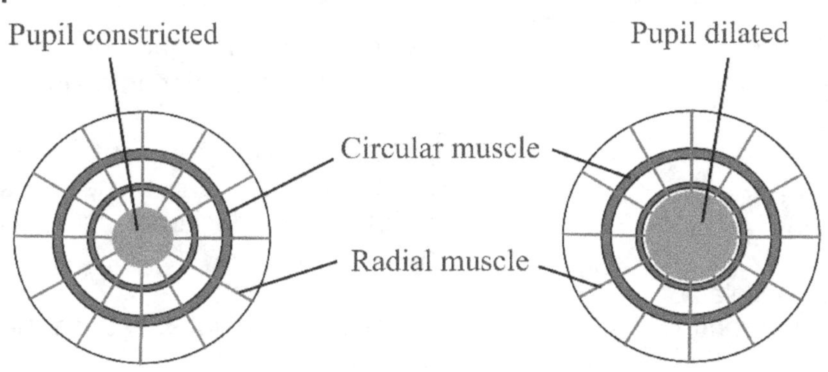

High light intensity
Circular muscles: contracted
Radial muscles: relaxed
Pupil diameter: small

Low light intensity
Circular muscles: relaxed
Radial muscles: contracted
Pupil diameter: large

The iris consists of radial and circular muscles, as shown in figure 11.4.

In bright light, the circular muscles of the iris contract and the radial muscles of iris relax. The pupil gets smaller (constricts) and less light enters the eye.

In dim light, the circular muscles of the iris relax and the radial muscles of iris contract. The pupil gets wider (dilates) and more light enters the eye.

Fig. 11.4: Constriction and dilation of pupil

Focussing images from near and distant objects on the retina

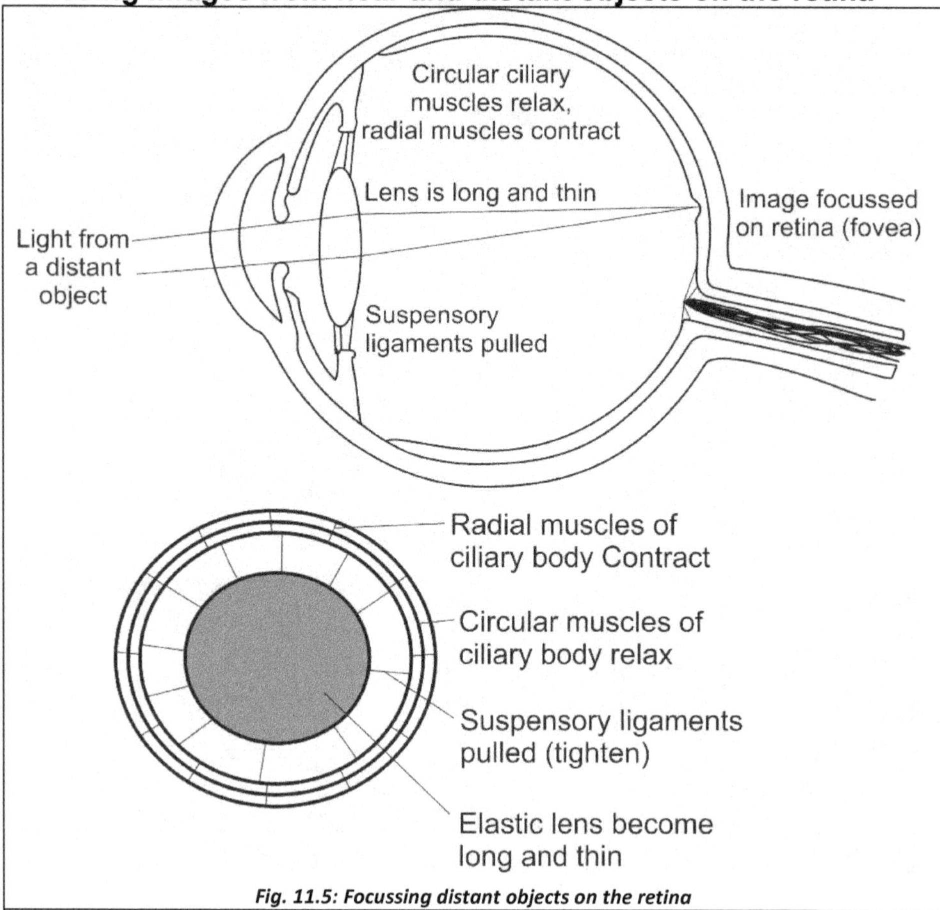

Fig. 11.5: Focussing distant objects on the retina

Distant object

The rays of light from a distant object are almost parallel and need to be bent only slightly by the lens to produce a focussed image on the retina, as shown in figure 11.5.

The circular ciliary muscles relax and radial ciliary muscles contract.

The Ciliary body pulls the suspensory ligaments and the suspensory ligaments become tense or taut.

The suspensory ligaments pull the lens and make it longer and thinner.

A clear image is focussed on the fovea (retina).

Fig. 11.5: Focussing near objects on the retina

Near object

The rays of light from a near object are almost diverging and need to be bent sharply by the lens to produce a focussed image on the retina, as shown in figure 11.6.

The circular ciliary muscles contract and radial ciliary muscles relax.

The suspensory ligaments are loosened and they become slack.

The suspensory ligaments stop pulling on the lens. The lens shortens and thickens as it is elastic.

A clear image is focussed on the fovea (retina).

Structure and functions of neurones

Neurones are nerve cells which carry electrochemical impulses from one part of the body to another.
Impulses always travel in one direction along a neurone – **from dendron** to **axon, across the cyton.**
The structure and function of the three basic types of neurones are shown and described below.

A. Sensory neurone: sensory neurones transmit **impulses from the sense organs to the central nervous system (brain or spinal cord).**

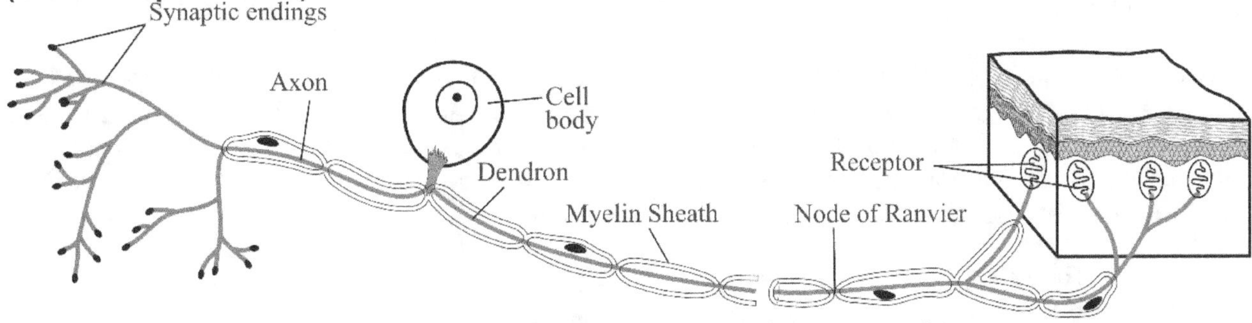

Fig. 11.6: Sensory neurone attached to a receptor cell

B. Motor neurone: motor neurones always transmit impulses **from** the **central nervous system** (brain or spinal cord) **to** the **effector organs** (muscles or glands).
There are many dendrons but a single long axon which is myelinated. They are sometimes referred to as multipolar neurones as they have many dendrites.

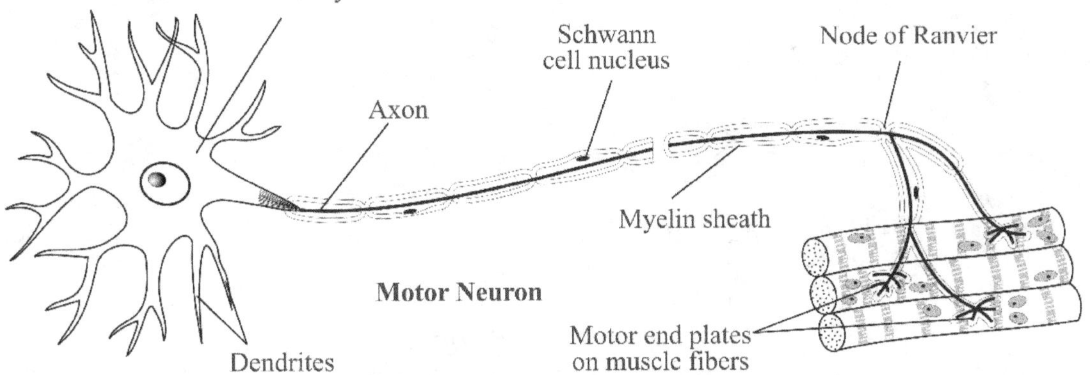

Fig. 11.7: Motor neurone attached to effector

C. Relay neurone: these are also called connector neurones or intermediate neurones, they are non-myelinated. The main function of these neurones is to transmit impulses from one neurone to another. In the brain and spinal cord a single relay neurone connects with many different neurones and helps to analyse information received from many neurones. The connections are made by **synapses**.

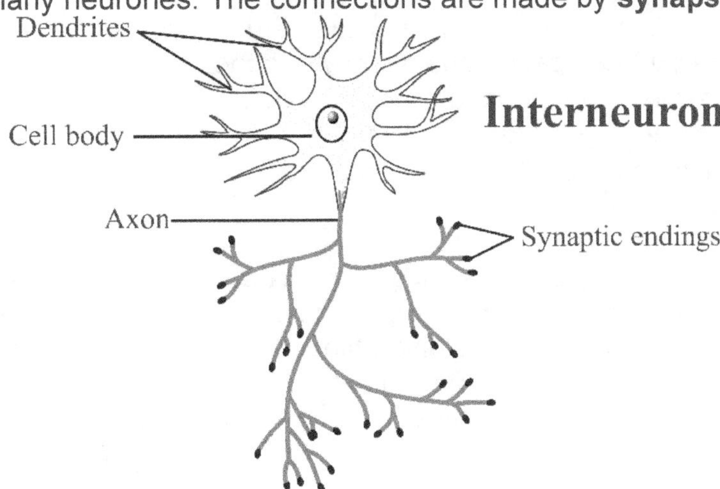

Fig. 11.8: Relay neurone

Figure 11.9 represents a simple spinal reflex. The path travelled by the impulses to bring about a reflex action is called a reflex arc. If the response is directed by the spinal cord then it is called a spinal reflex. A typical spinal reflex arc passes along the following path:

STIMULUS → RECEPTOR → SENSORY → RELAY → MOTOR → EFFECTOR → RESPONSE
(CHANGE IN (SENSE ORGAN) NEURONE NEURONE NEURONE (MUSCLE (CHANGE IN BEHAVIOUR /
ENVIRONMENT) OR GLAND) PHYSIOLOGY)

Spinal reflexes are useful to allow animals to respond quickly to danger. The impulse to operate the effector organs is generated by the spinal cord itself. This helps to save valuable time for the organism to respond. This ability to respond quickly to a stimulus increases an organism's chances of survival. This response is usually an unconscious response, without the involvement of the brain. However, the impulse will be sent to the brain later on and the organism becomes conscious of the response.

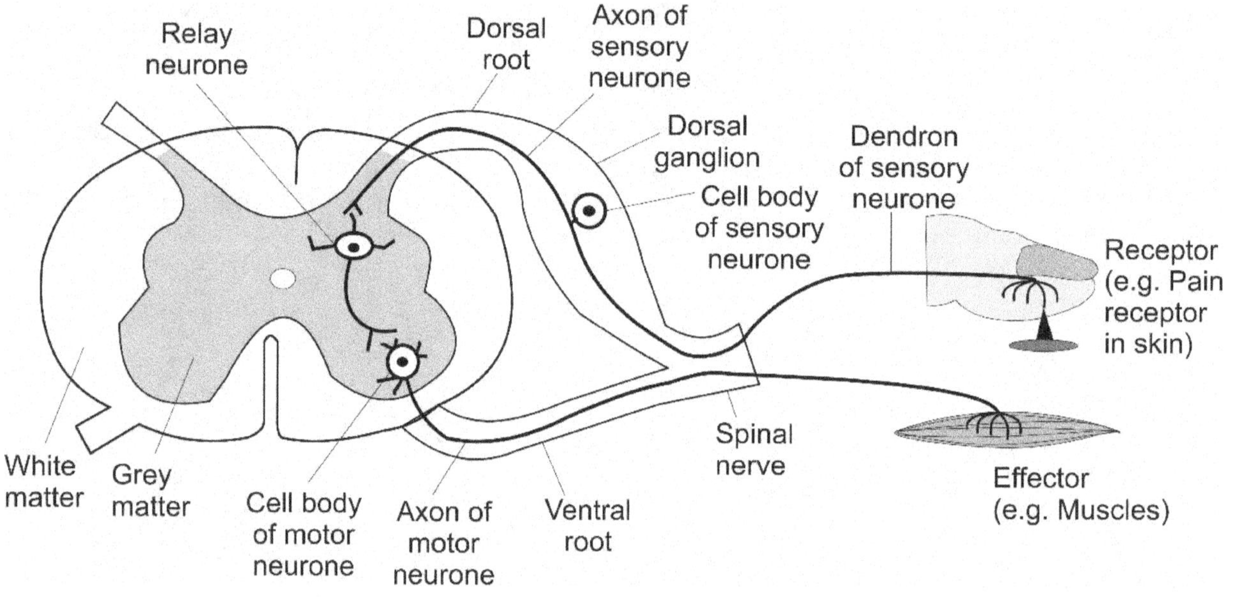

Fig. 11.9: A simple spinal reflex

- Hormones are chemical messengers which are secreted by endocrine glands. These glands are ductless glands which pour their secretions directly into the blood.
- The hormones are carried by the blood to all parts of the body, but they affect only specific organs. These organs are called as target organs.
- The hormones alter metabolic activity in the target organs.
- The liver will then destroy the hormones.

For example, adrenaline is secreted by the adrenal gland, just above the kidneys. It is carried by the blood to all parts of the body, but acts upon the liver and muscle cells. It increases the blood glucose levels and can increase the heartbeat rate, dilate the pupils, and divert blood from the periphery of the body to vital organs like brain and liver. This helps the person to face an emergency. The muscles will not tire easily due to rapid supply of oxygen and glucose, due to increase in the rate of aerobic respiration within the muscle cells.

Insulin is secreted by the Beta cells in the Islets of Langerhans in the pancreas. It is carried by the blood to all parts of the body but influences only liver and muscle cells. The action of Insulin is described below.

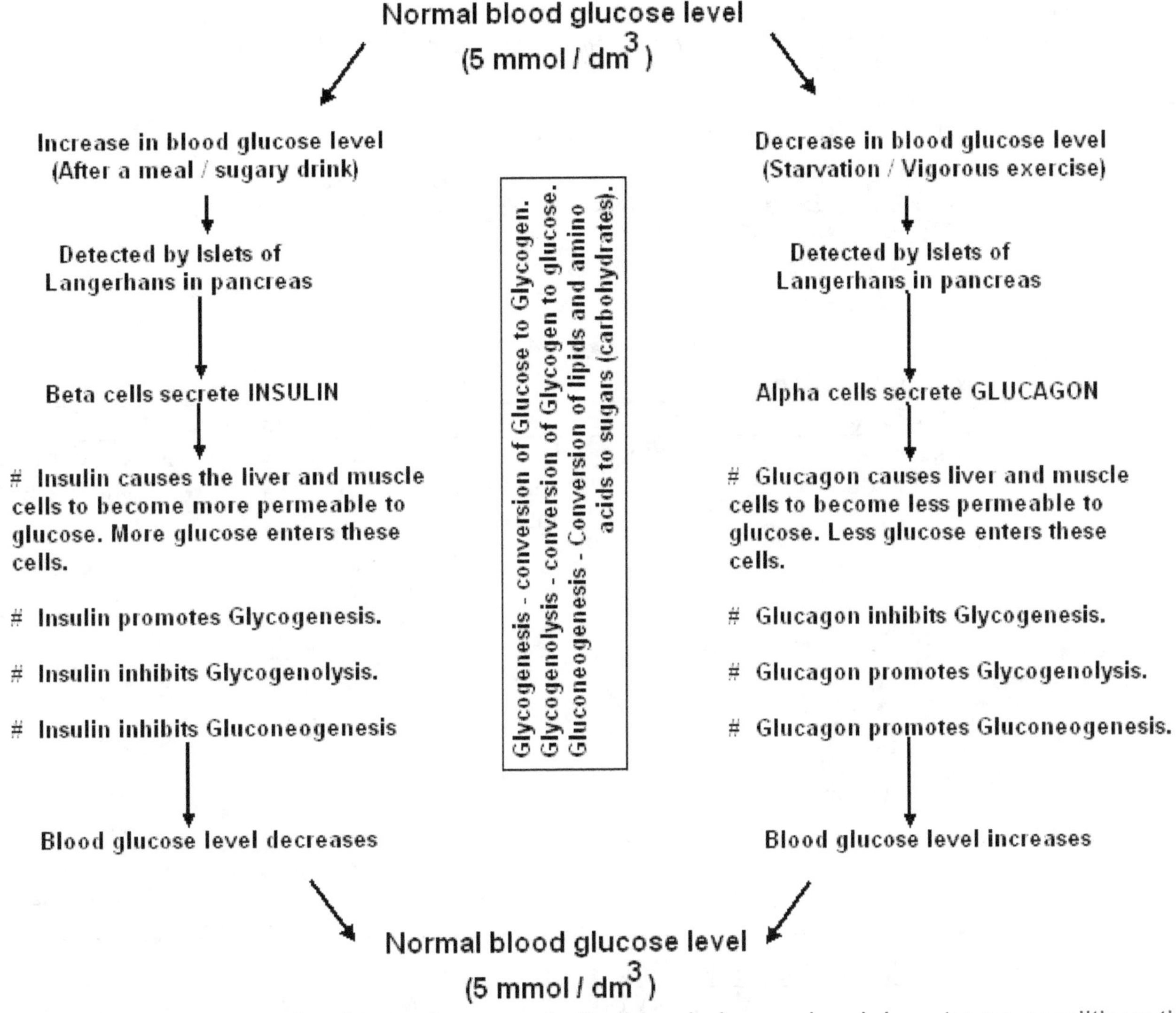

Lack of insulin in the blood can lead to an increase in the blood glucose level. In extreme conditions the glucose may be excreted in the urine. This is called as diabetes mellitus. It can be treated by injecting insulin into the blood or by inhaling it into the lungs.

Differences between nervous and hormonal control

Nervous control	Hormonal control
Transmitted by electrochemical impulses	Transmitted by chemical messengers (hormones)
Speed of transmission very quick	Speed of transmission much slower
The effect is usually very short lived	The effect is usually long lasting
Transmitted through neurones	Transmitted through blood
Usually affects a specific organ	Usually has a widespread effect and can affect many organs.

> *Cambridge 5090 syllabus specification 12(a)* identify and describe, from diagrams, photographs and real specimens, the main bones of the forelimb (humerus, radius, ulna and scapula) of a mammal;
> *Cambridge 5090 syllabus specification 12(b)* describe the type of movement permitted by the ball and socket joint and the hinge joint of the forelimb;
> *Cambridge 5090 syllabus specification 12(c)* describe the action of the antagonistic muscles at the hinge joint.

Bones of the forelimb of a mammal

Figure 12.1 shows the main bones of the forelimb of a mammal.

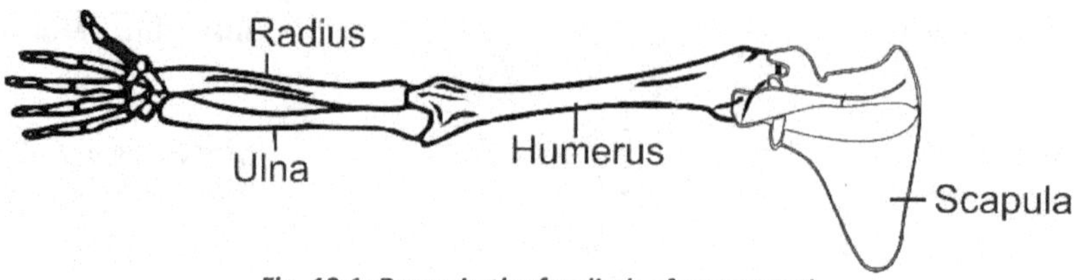

Fig. 12.1: Bones in the forelimb of a mammal

Movement at elbow and shoulder joint

Elbow joint	Shoulder joint
The elbow joint allows the bones to move by 180^0 in one plane only, as shown in figure 12.2.	The shoulder joint allows rotatory movement (360^0) and the bones can move in two planes, as shown in figure 12.3.

Fig. 12.2: Elbow joint

Fig. 12.3: Shoulder joint

Antagonistic action of muscles at the hinge joint

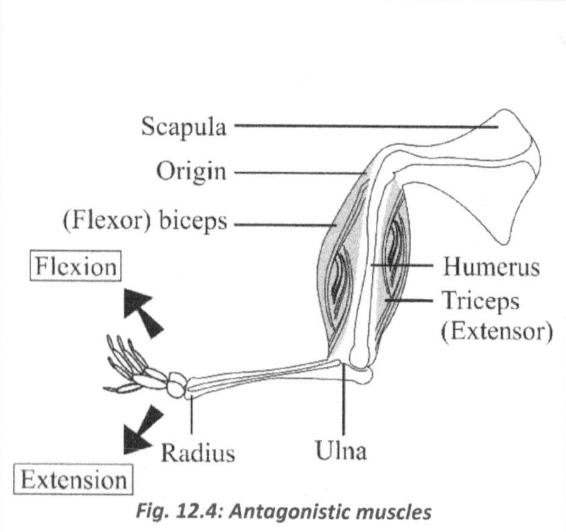

Fig. 12.4: Antagonistic muscles

Muscles work in pairs. As one contracts the other in the pair relaxes. One muscle produces the opposite movement from the other muscle, therefore, the pairs are called **antagonistic pairs**.

Muscles which cause a joint to extend or straighten are called **extensors**, muscles which cause a limb to retract, bend or fold are called **flexors**. **Flexion** is the movement that reduces the angle of a joint. **Extension** increases the angle of a joint and moves the bones away from each other.

The biceps are attached to the scapula at one end and to the radius at the other end. So, when the bicep contracts the arm is bent at the elbow. This is called as flexing and the bicep is the flexor.

To straighten the arm or extend the arm, the triceps contract and pull the ulna. So, the triceps act as extensors, as shown in figure 12.4.

Cambridge 5090 syllabus specification 13(a) define a drug as any externally administered substance that modifies or affects chemical reactions in the body;
Cambridge 5090 syllabus specification 13(b) describe the medicinal use of antibiotics for the treatment of bacterial infection;

A drug is an externally administered chemical substance which modifies or alters metabolic reactions in the body.

Antibiotics are useful drugs which are used to kill or to slow down the growth of bacteria in the body. They do not harm human cells but can harm bacterial cells. Antibiotics cannot be used to treat infections caused by viruses. It is used only for bacterial infections, like tuberculosis and syphilis.

Analgesics are drugs which are used as pain killers. Antibiotics and analgesics are examples of useful drugs, called as medicinal or pharmaceutical drugs.

Heroin, alcohol and nicotine are harmful narcotics. These are drugs which affect your mood and behaviour and are usually addictive.

Investigating the effect of antibiotics on bacteria.
Step one: Preparation of nutrient medium
A nutrient medium is prepared and sterilised. The nutrient medium is poured into a Petri dish and allowed to solidify, as shown in figure 13.1.

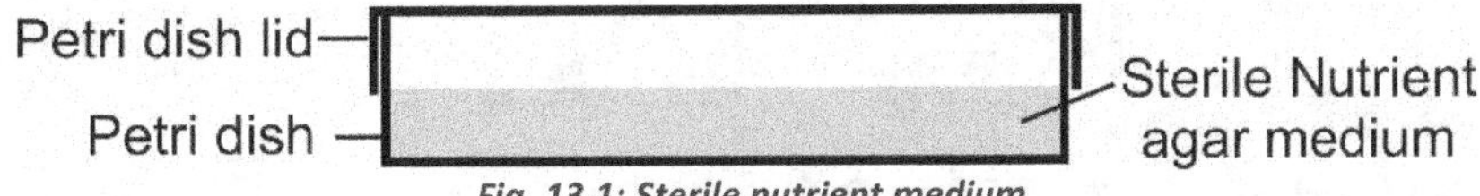

Fig. 13.1: Sterile nutrient medium

Step two: Inoculation of medium.	Step three: The bacterial culture is spread uniformly on the surface of the agar.
Introducing the bacteria (inoculum) into the nutrient medium is called as inoculation. Use a sterile pipette to transfer a sample of the inoculum on to the surface of the nutrient agar in the Petri dish. **Note:** a non-pathogenic strain of bacteria is suitable for culturing, as it reduces the risk of infections and is not hazardous to human health. K12 strain of *E.coli* or bacteria from yoghurt or soil may be chosen.	Use a sterile spreader to spread the inoculum over the agar surface, as shown in figure 13.2. 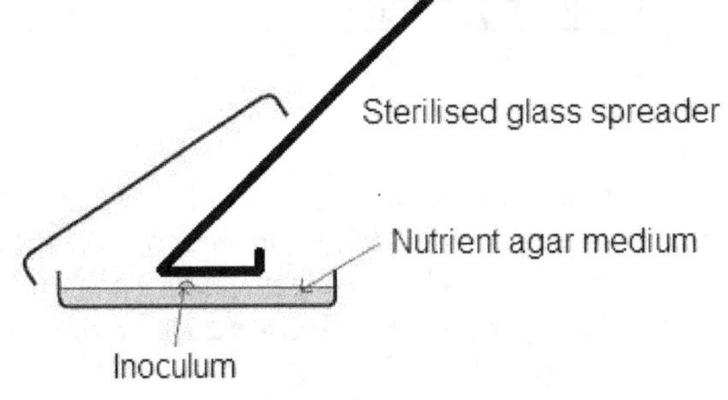 *Fig. 13.2: Spreading the bacteria on the surface of agar*

Step four: Incubate the bacteria with the antibiotics.	Step five: Observe the area of the clear zone
Prepare a solution of the antibiotic and soak a filter paper disc in the antibiotic. Place the disc onto the **surface of the inoculated agar medium**. A control can be a disc soaked in distilled water. Incubate the dish at 25⁰C for 24 hours.	The antibiotic prevents bacteria from growing around the disc. Bacteria grow in all other regions of the dish. 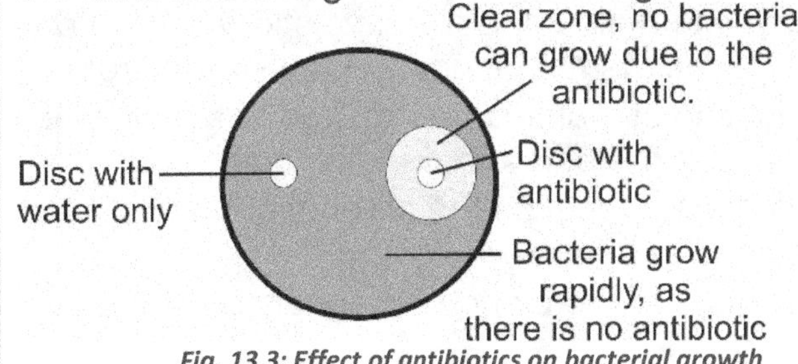 *Fig. 13.3: Effect of antibiotics on bacterial growth*

Heroin

- Heroin is a depressant which slows down the functioning of the nervous system.
- It is an addictive drug, which means that the person starts to crave or have an unstoppable desire to keep using the drug.
- If the use of the drug is stopped then there are severe withdrawal symptoms, like headaches, nausea, nervousness, etc. This prevents the users from discontinuing the use of the drug.
- As the person keeps using the drug, the body starts to develop tolerance to the drug and the person must take higher doses to get the same effect.
- Heroin is injected into the blood stream (veins). Many users may share the same needle (due to economic difficulties) to inject the drug. This increases the risk of transmission of AIDS and Hepatitis.

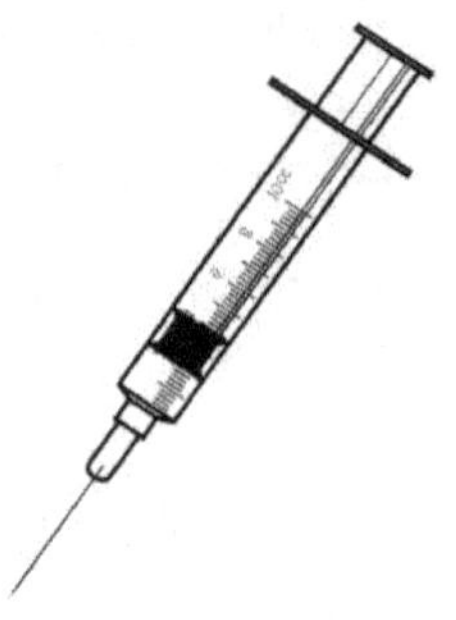

Fig. 13.4: Syringe

Alcohol

- Alcohol is also a depressant which slows down the functioning of the nervous system.
- It increases the reaction time and slows down nervous responses. That is why drinking and driving is illegal. Drivers will be slow to respond to any obstacles and the risk of accidents in much higher.
- Alcohol consumption also reduces self-control and removes social inhibitions.
- Alcohol is removed from the blood and broken down by the liver. Excess consumption of alcohol can cause liver damage or cirrhosis.
- Alcohol is very expensive and leads to social problems stemming out of excess spending. Many families face economic difficulties when alcoholism prevails in the family.

Fig. 13.5: Alcoholic beverage

Nicotine

- Nicotine stimulates the brain to release chemicals (dopamine) from the brain. These chemicals give a smoker a false sense of wellbeing. This makes nicotine an addictive drug which prevents people from giving up smoking.
- Nicotine makes the platelets sticky. This increases the risk of blood clotting inside the arteries and can increase the risk of heart attacks and stroke.
- Nicotine stimulates the release of adrenaline. This increases the heartbeat rate and blood pressure. The risk of heart attacks also increases.
- Nicotine increases the amount of harmful fats in the blood. This increases the risk of atheroma formation in arteries.
- Nicotine in the blood of pregnant mothers can result in low birth weight of the foetus. must be with ref. nicotine)

Tar (A mixture of many chemicals in cigarette smoke)

Tar is sticky and brown, and stains teeth, fingernails and lung tissue. Tar coats the cilia in the trachea, causing them to stop working and eventually die. This reducing the filtering of air and allows solid particles in tar to enter the alveoli. Many of these chemicals in tar may increase the risk of lung cancer.

Carbon monoxide

Carbon monoxide binds to oxygen to form carboxyhaemoglobin. The carboxyhaemoglobin prevents haemoglobin from binding with oxygen and decreases the oxygen carrying capacity of the blood.

Emphysema

Emphysema is the effect on the alveoli in the lungs due to long term smoking.

- The alveoli walls breakdown, reducing the surface area for gas exchange.

- The capillaries in the lungs are reduced. So, surface for gas exchange is further reduced.

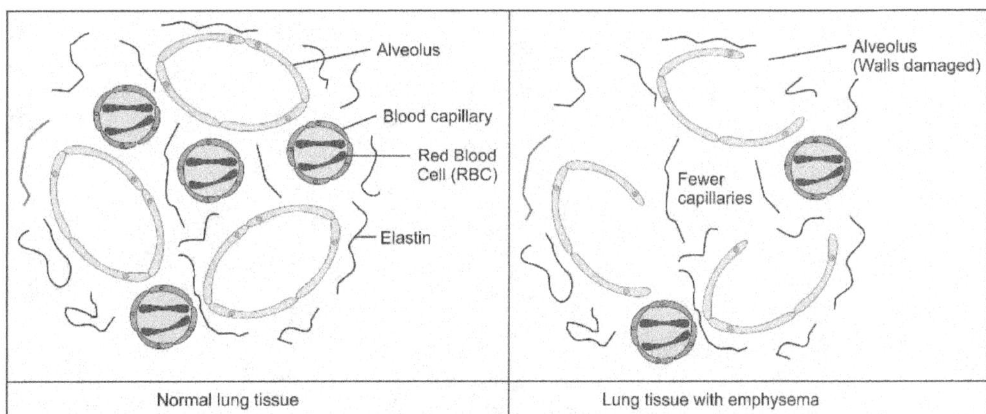

Fig. 13.6: Emphysema

Passive smoking

The signboards in figure 13.7 show that smoking is no longer acceptable and desirable by the general public. Special designated areas for smoking in public places may be used by smokers. In many countries smoking in public is a punishable offence.

The reason for this is that non-smokers may be exposed to the risks of passive smoking. Many people are nauseated by the smell of cigarette smoke. It also may cause irritation and burning of eyes in some non-smokers. Second hand smoke increases the risk of lung cancer and heart diseases.

Fig. 13.7: Smoking in public places

Cambridge 5090 syllabus specification 14(a) list the main characteristics of the following groups: viruses, bacteria and fungi;

Bacteria

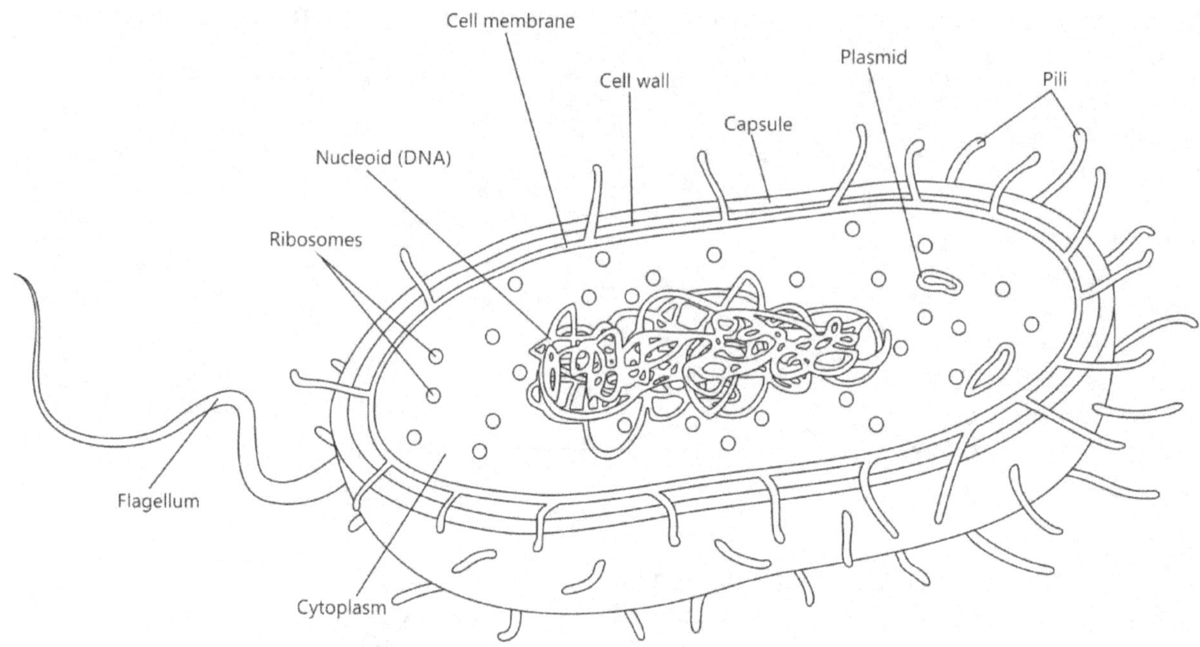

Fig 14.1 Bacteria structure (Internal)

Some of the characteristic features of bacteria are given below.

- Bacteria are single celled organisms that do not have a true nucleus. The DNA lies in the cytoplasm of the cell.
- Bacteria has circular DNA.
- Bacteria have a cell wall made up of murein or peptidoglycan.
- Plasmids are small circlets of DNA found in bacterial cells.
- Bacteria secrete enzymes to digest their food. The digested food is then absorbed into the cells. this is called extra-cellular digestion.

A typical virus - Viruses are Non-Cellular bundles of nucleic acids and proteins.

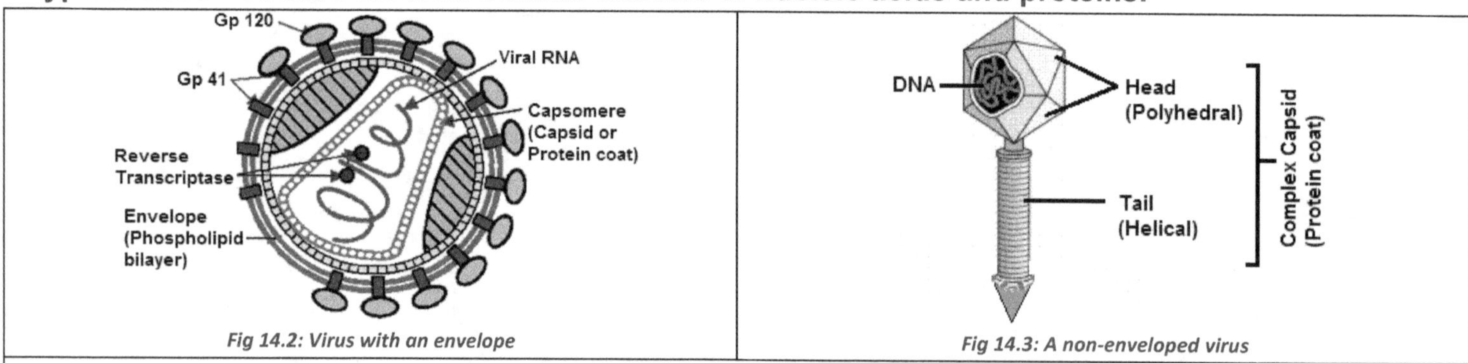

Fig 14.2: Virus with an envelope	*Fig 14.3: A non-enveloped virus*

- All viruses have a protein coat (the capsid), which contains genetic material. The genetic material is either DNA or RNA (Never both).
- The protein capsid is made from identical subunits (called capsomeres). The capsomeres can be arranged into variable shapes. In addition, some viruses also have an outer envelope surrounding the capsid.
- All viruses are parasitic. They multiply inside the cells Biof the host.

Fungi

Fungi are a group of organisms which may be single celled, like yeast, or multicellular, like moulds. Some features of fungi are:

- non photosynthetic eukaryotic organisms, with a non cellulose cell wall;
- absorptive methods of nutrition;
- usually have multinucleate hyphae;
- form spores without flagella.

e.g. Yeast and moulds

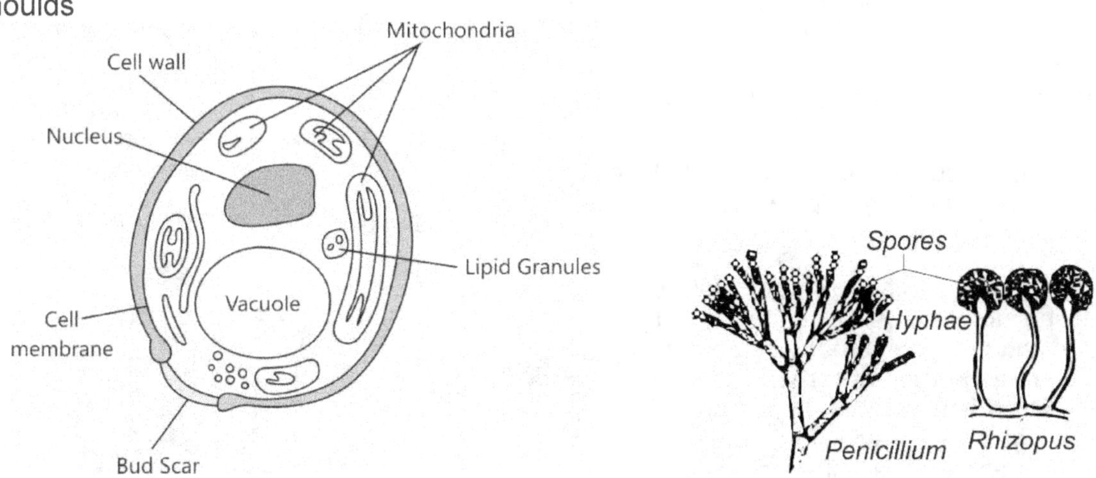

Fig. 14.4: Yeast cell

Fig.14.5: Filamentous fungi (Penicillum and Rhizopus)

Cambridge 5090 syllabus specification 14(b) outline the role of microorganisms in decomposition;

Role of Micro-organisms in decomposition and recycling of carbon

Decomposition is the natural process carried out by bacteria and fungi. The organic compounds in dead animal or plant tissue are converted into simple inorganic compounds that can be reused by plants and animals. The conversion of some simple molecules is shown in fig 14.6.

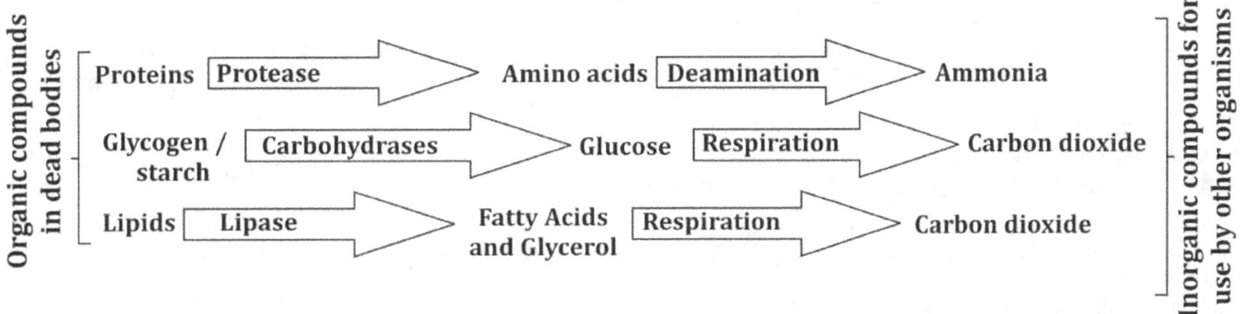

Fig.14.6: Role of microorganisms in decomposition
(Converting organic compounds into inorganic compounds)

If dead bodies and excretory products of plants and animals did not decompose, the Earth's surface would soon be covered in a deep layer of undecomposed organic matter. Fortunately this does not happen because dead organisms and animal wastes become food for some other organisms (detrivores and decomposers) to live on.

During the process of decomposition, the decomposers obtain food for themselves by extracting chemicals from the dead bodies or organic wastes and use these chemicals to produce energy. Other chemicals will return to the soil and be reused by plants. Death and decomposition are an essential part of all life cycles on earth. To enable successful birth and growth of young plants and animals, older specimens must die and decompose. This limits the competition for resources and provides a fresh source of essential nutrients for new generations of life.

Bread production

Flour (100 parts) + Water (60 parts) + Yeast (5 parts)
+ sugar (1 part) + Salt (1 part)

↓ Low speed mixing for 15 minutes

Dough left to stand for 2 hours at 30^0C

↓

Glucose $\xrightarrow{\text{Yeast}}$ Ethanol + Carbon dioxide

↓

Dough rises due to carbon dioxide bubbles
being trapped in the spaces within the dough

↓

Heat at 230^0C till a brown crust forms
Heating kills yeast, causes the bread to
rise and become porous and spongy
due to expansion of carbon dioxide
bubbles and allows the ethanol to
evaporate, leaving behind a
characteristic flavour.

Alcohol production

Prepare a fruit or malt extract.
Sugars are present in the extract.

↓ Boil to sterilise and remove oxygen

Cool to 30^0C and inoculate with yeast

↓ Incubate at 30^0C
and maintain
anaerobic
conditions

Glucose $\xrightarrow{\text{Yeast}}$ Ethanol + Carbon dioxide

Yoghurt production

Milk (contains lactose, proteins and lipids)

Heat at 90^0C for 20 minutes to
- kill bacteria
- remove air, favouring the growth
  of micro-aerophillic lactobacillus
- denature some proteins to give
  the yoghurt a thicker consistency

Homogenise the milk by using mixers or tuning forks to
- disperse fat globules. This makes the yoghurt creamier
and less likely to separate into cream and whey later on.

Cool the milk to 40^0C to prevent the
death of bacteria in the starter culture

Inoculate the milk with the starter culture of
Lactobacillus bulgaricus and *Streptococcus thermophilus*

↓

Incubate at 40^0C for 6 hours until the pH reaches 4.5

Chemical changes during incubation

Milk proteins ⟶ Peptides and amino acids

Lactose ⟶ Formic acid

Lactose ⟶ Lactic acid + Ethanal

pH of 4.5 stops further action of bacteria and
marks the **end of fermentation**

Cool the yoghurt to 4^0**C**, pack and distribute

Cheese production

Milk (contains lactose, proteins and lipids)

Heat at 50^0C for 15 minutes
to kill bacteria

↓

Cool the milk to 35^0C, add rennet (chymosin)
and *Lactobacillus (Bacteria)*

↓

Lactose $\xrightarrow{\text{\textit{Lactobacillus (Bacteria)}}}$ Lactic acid
(Low pH)

↓

Activation of chymosin
due to low optimum pH

Casein protein
in milk solidifies
to form **curd**   **A liquid whey**
forms

↓

Salting and further
fermentation (ripening)
of the curd by action of
bacteria and fungi to give
a characteristic flavour to
the cheese.

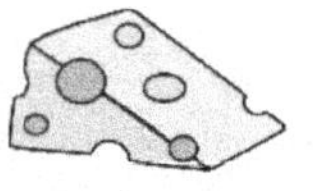

Penicillin production in fermenters

- Penicillin is produced by growing the fungus *Penicillium chrysogenum* in large scale fermenters with a capacity of up to 200000dm^3.
- Steam is passed through the empty fermenters under high pressure to sterilise the vessel. The nutrients medium is sterilised by ultraheat treatment or UV radiation. This prevents contamination of the medium and does not allow other microorganisms to grow in the fermenter.
- The commonly used nutrients are lactose or glucose as a carbohydrate source.
- The nutrient medium inoculated with a pure strain of *P. chrysogenum.*
- Sterile air (filtered / heated and cooled) is pumped through the medium through the sparger. The impeller stirs the medium increasing the rate at which oxygen dissolves. It also prevents clumping of mycelia and promotes efficient heat exchange between media and cooling surfaces.
- The acid base reservoir can be used to regulate the pH between a range of 6.8 to 7.7. The cooling jacket regulates temperature at 25^0 C to 27^0 C.
- Penicillin is produced by the fungus and secreted into the medium in the fermenter. The process is usually completed in about 6 to 7 days.
- The penicillin is separated from the other components of the fermenter and used as an antibiotic.

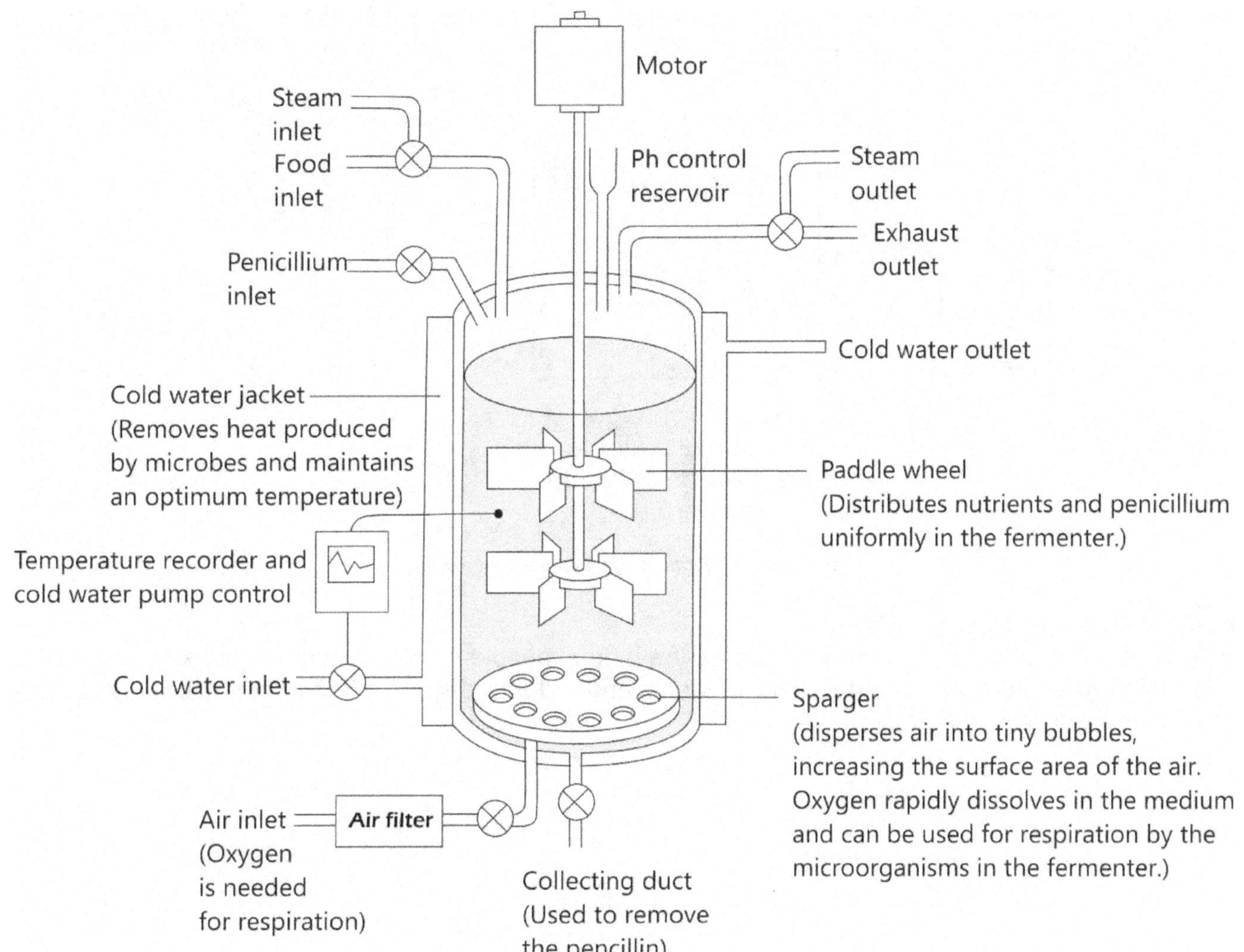

Fig.14.7: A large scale fermenter for penicillin production

Single cell protein production in large scale fermenters
Single celled proteins is obtained by growing fungi, algae or bacteria to produce edible food with a high concentration of protein.

The cells are grown as a continuous culture in large fermenters.

The sterile nutrient broth contains:
Glucose: which is a carbon and energy source for the culture.

Growth factors to increase length and diameter of fungal hyphae.

Compressed air: Provides oxygen for respiration. It also agitates the culture medium preventing nutrients and microbes from settling down. This type of fermenter is called an 'Air Lift' fermenter.

Ammonia: It is a source of nitrogen for protein synthesis. It also maintains the pH of the culture medium between 6 to 6.7, by neutralizing carbonic acid and other organic acids produced by metabolism.

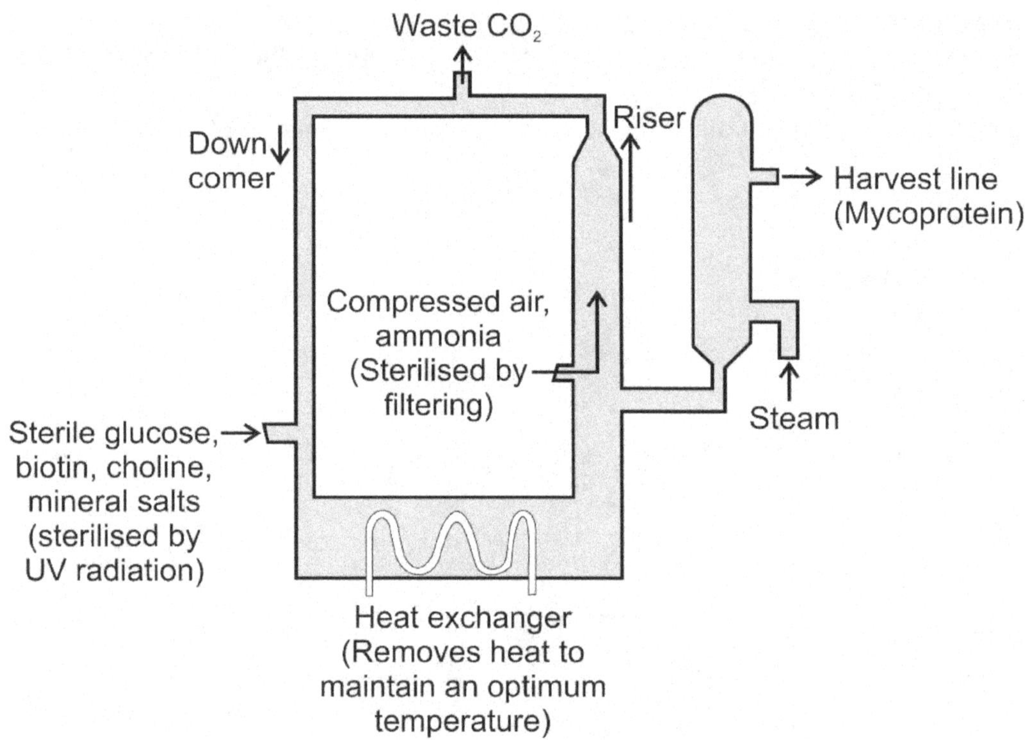

Fig.14.8: An air-lift fermenter for producing mycoprotein

Advantages of single celled proteins
Single cell proteins are useful as a food source as the land available for agriculture is shrinking. Moreover, the production of protein rich food in fermenters is faster and independent of climatic conditions as opposed to traditional food produced by plants.

Cambridge 5090 syllabus specification 15(a) state that the Sun is the principal source of energy input to biological systems;
Cambridge 5090 syllabus specification 15(b) describe the non-cyclical nature of energy flow;

Solar energy is the primary source of energy for all ecosystems. The autotrophs use solar energy to prepare food by photosynthesis. The organic compounds formed by plants are then a source of food for all other organisms in the ecosystem. The **energy is finally lost as heat to the surrounding**. So, to keep an ecosystem functioning efficiently, solar energy must constantly be supplied. Figure 15.1 shows the flow of energy through the various components of an ecosystem.

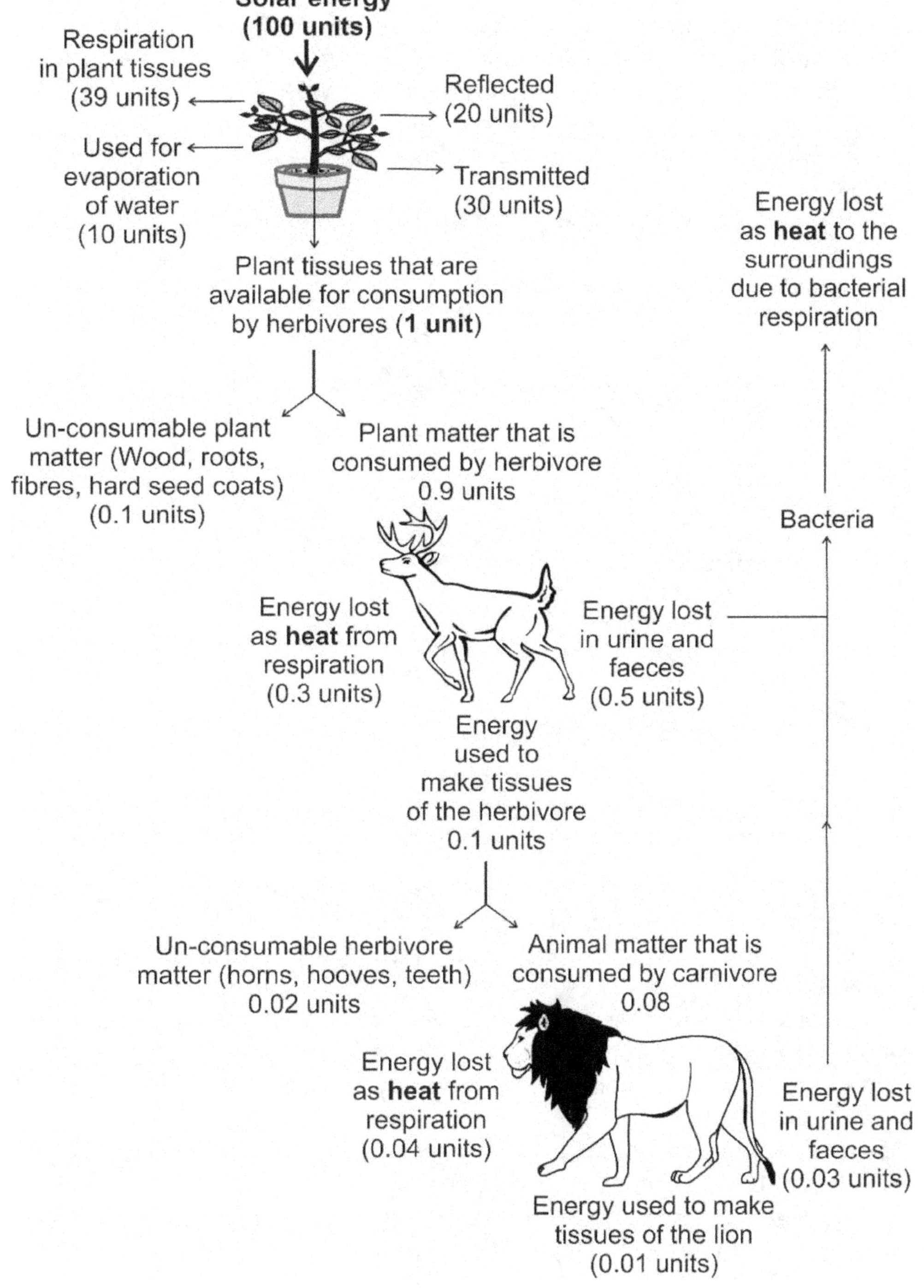

Fig. 15.1: Non-cyclical nature of energy flow through an ecosystem

Producer – an organism that makes its own organic nutrients, usually using energy from sunlight through photosynthesis. Example, all green plants.

Consumer – an organism that gets its energy by feeding on other organisms. Example, caterpillars, birds and lions.

Herbivore – an animal that obtains its energy by eating plants. Example, cow or deer.

Carnivore – an animal that obtains its energy by eating other animals. Example, lions and wolves.

Decomposer – an organism that obtains its energy from dead or waste organic matter. Example, bacteria and fungi.

Food chain – a chart showing the flow of energy (food) from one organism to the next, beginning with the producer.

e.g. mahogany tree → caterpillar → songbird → hawk

Food web – a chart showing many interconnected food chains in the ecosystem, as shown in figure 15.2.

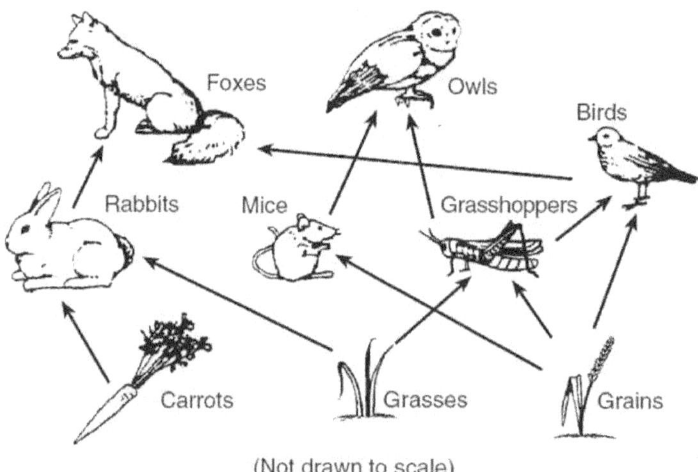

(Not drawn to scale)

Fig. 15.2: A typical grassland food web

The energy from plant biomass is transferred from one trophic level to the next in the form of organic compounds (food) along the food chains. There is a lot of energy lost between the trophic levels. The main reasons for energy losses are:

- Not all parts of the plants or animals are consumed by the next trophic level. The roots bones, teeth, claws are not always eaten.
- All the consumed food is not assimilated into the body. A large part remains undigested and is lost in faeces (E_F)
- A lot of the assimilated food is used up for respiration (E_R).
- Some energy is lost in urine (E_U) as chemical energy.
- Only about 10% of the food consumed is incorporated into biomass (E_P) of the consumer.
- The fate of energy consumed (E_C) by herbivores can be summarized by the equation.

$$E_C = E_P + E_R + E_U + E_F$$

Consider the food chain in figure 15.3. Almost 90 percent of the energy is lost during transfer from one trophic level to another. So, the top carnivore has a very small, inadequate amount of energy available for another trophic level to consume. This is why food chains are usually very short.

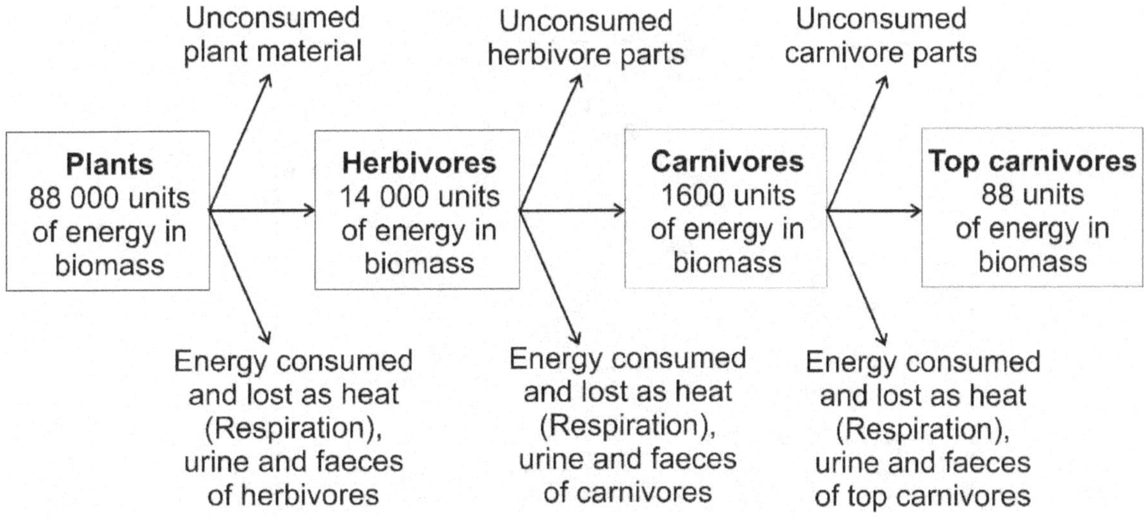

Fig. 15.3: Energy losses limit the length of food chains

Pyramid of numbers – It is a graphical representation of the number of organisms at each trophic level. The number of producers is represented at the bottom and the numbers of organisms at higher trophic level are placed centrally above the lower trophic levels, as shown in A, B and C in figure 15.4. The same scale must be used to represent the width of each trophic level. The disadvantage is that the number of organisms at each trophic level is not always proportionate to the energy content. Thus some pyramids are inverted, as in example B and C in figure 15.4.

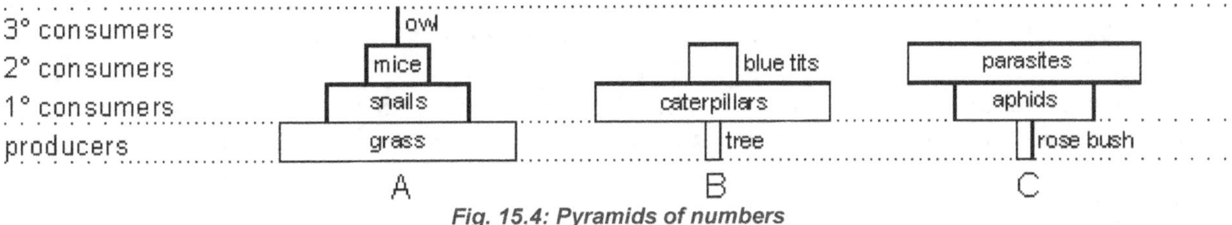

Fig. 15.4: Pyramids of numbers

Pyramid of biomass - this is graphical representation (drawn to scale) showing the biomass of each trophic level. Biomass is defined as the mass of organisms per unit area of ground at any given time (standing crop mass). These pyramids are usually upright, but sometimes may be inverted. Biomass is measured as Kg/m^2.

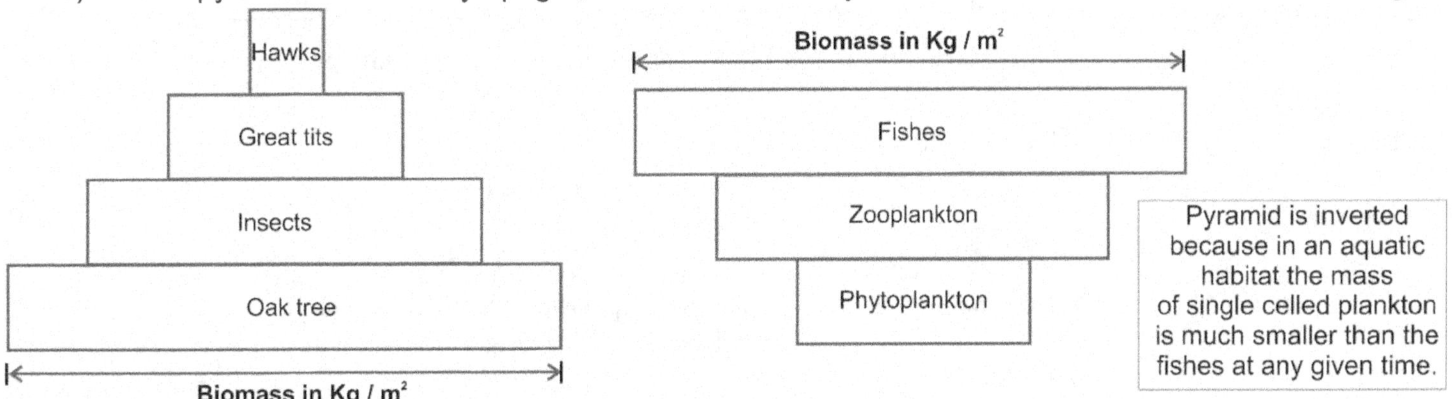

Fig. 15.5: Pyramids of biomass

Cambridge 5090 syllabus specification 15(f) describe and state the importance of the carbon cycle;

Figure 15.6 shows the carbon cycle. The processes in the carbon cycle regulate the carbon dioxide concentration in the atmosphere.

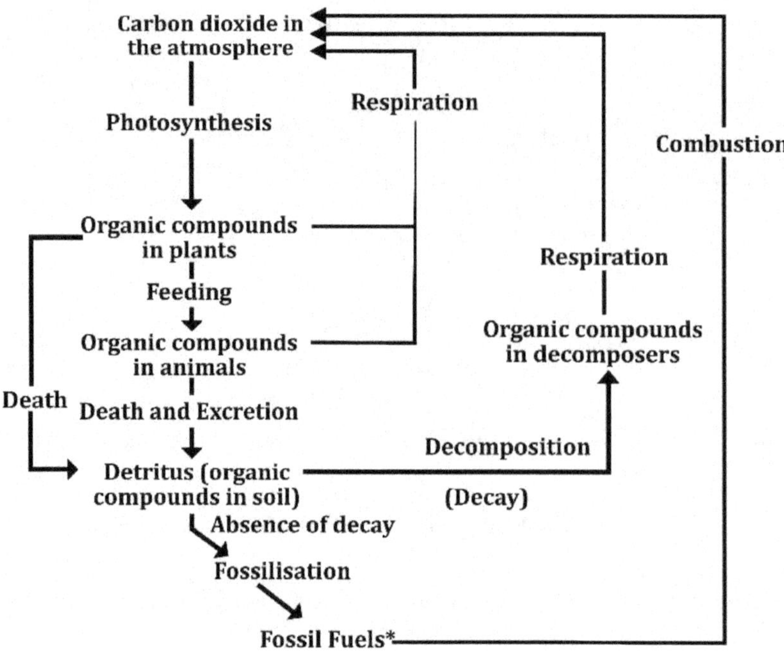

Fig. 15.6: Carbon cycle

- Photosynthesis is the only process that removes carbon dioxide from the atmosphere and converts it into complex organic compounds. The carbon dioxide is fixed into the biomass of the producers.
- Respiration and combustion are processes that release carbon dioxide into the atmosphere.
- Carbon sinks trap carbon in the form of carbonates, wood of trees, fossil fuels and non-biodegradable plastics for long periods of time. Destruction of carbon sinks by human activities - deforestation, burning of fossil fuels, burning of corals - has led to an increase in the level of carbon dioxide in the atmosphere, as carbon dioxide is released into the atmosphere faster than it can be removed.

Cambridge 5090 syllabus specification 15(g) describe the nitrogen cycle in making available nitrogen for plant and animal protein, including the role of bacteria in nitrogen fixation, decomposition and nitrification (details of denitrification and the names of individual bacteria are not required);

Note: The names of bacteria are not required for the exam. Just given to quell the curiosity of some students.

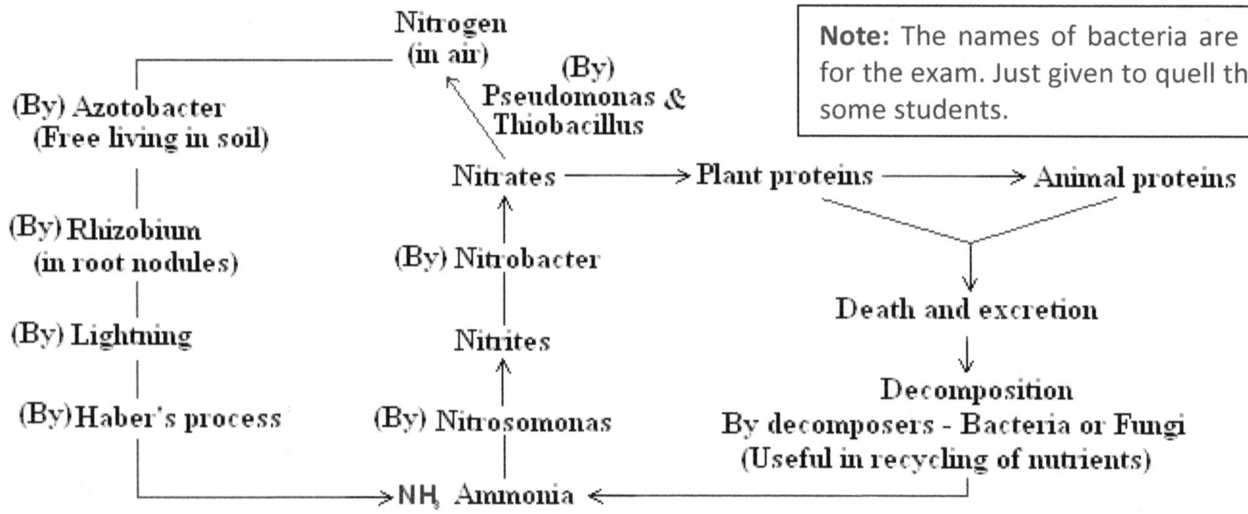

Fig. 15.7: Nitrogen cycle

- **Nitrogen fixation** - Conversion of nitrogen gas into soluble Nitrogenous compounds like ammonia, which can be used by plants.

- **Nitrification** - This is the conversion of ammonia to nitrites and nitrates. This process is enhanced by the presence of oxygen in the soil - well aerated, well drained, loose soil favours oxidation of ammonia. Tilling and ploughing of the soil helps to aerate the soil and maintain aerobic conditions.

- **Denitrification** - Conversion of Nitrites and Nitrates to nitrogen gas. This occurs in anaerobic conditions like water logged soil, clayey and compacted soil.

- **Decomposition -** Microbial saprophytes break down proteins in detritus to form ammonia in two stages: first they digest proteins to amino acids using extracellular protease enzymes, then they remove the amino groups from amino acids using deaminase enzymes.

Malaria is caused by a single celled organism called *Plasmodium*. The pathogen (*Plasmodium*) is carried from the blood of an infected person and injected into the blood of a healthy person by the female *Anopheles* mosquito. The mosquito is therefore called as a vector of the malarial parasite.

A vector is an organism that carries a parasite from one host to another.

The mosquito is an effective vector for the following reasons.
- The mosquito is a human parasite as it feeds on human blood.
- The mosquito has a needle-like mouth part which can pierce the skin and enter blood vessels. This helps to inject the Plasmodium from the salivary gland of the mosquito into human blood.
- The mosquito is attracted to warm blooded organism, like humans. It feeds at night when humans are asleep.
- The mosquitos breed very rapidly and are found in large numbers.
- The mosquito has wings which help it to fly from one host to another.

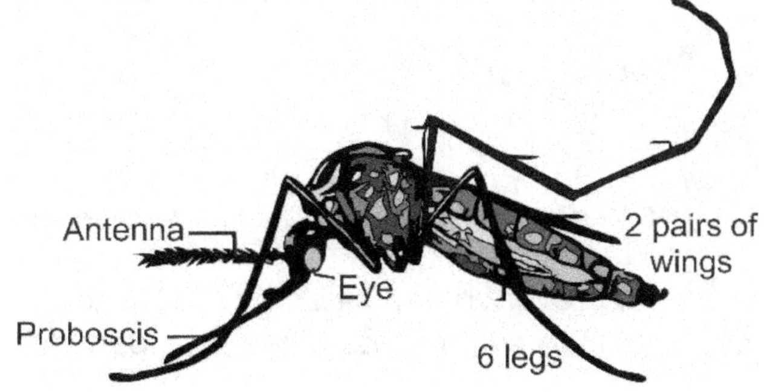

Fig. 15.8: Female Anopheles mosquito (Vector for malarial parasite)

- The mosquito releases an anticoagulant into the blood of humans. This prevents blood clotting and helps the blood to flow into the mosquitos stomach.
- The plasmodium multiplies in the body of the mosquito but does not kill the mosquito.

Measures that can be taken to prevent the spread of malaria
Prevent breeding of mosquitos
- This can be achieved by draining away stagnant water, which is used by mosquitos to lay their eggs. The larvae and pupa also grow and develop in stagnant water, as shown in figure 15.9.

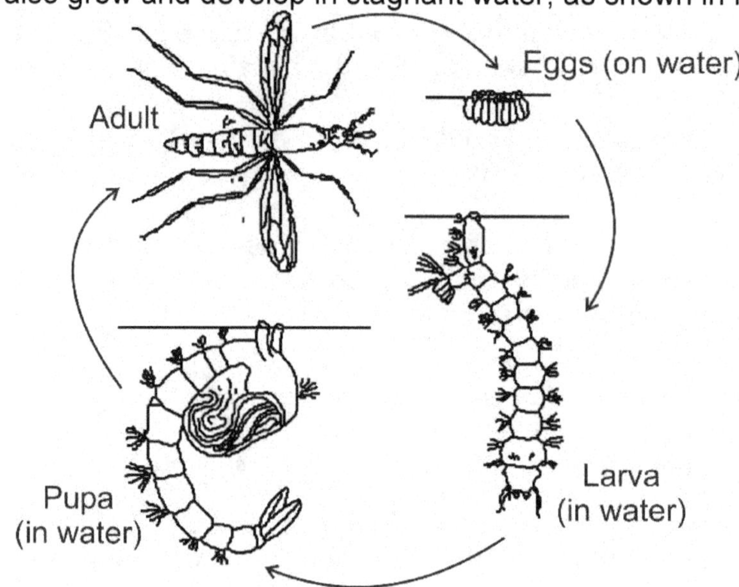

Fig. 15.9: Life cycle of mosquito

- Spraying a thin layer of oil on the surface of stagnant water will help to kill the larvae as they cannot respire.
- Guppy fish feed on mosquito larvae and help to control mosquito populations.

Bite prevention
- Sleep under mosquito nets to avoid getting bitten at night. Fix nets on windows to prevent the entry of mosquitos into the room.

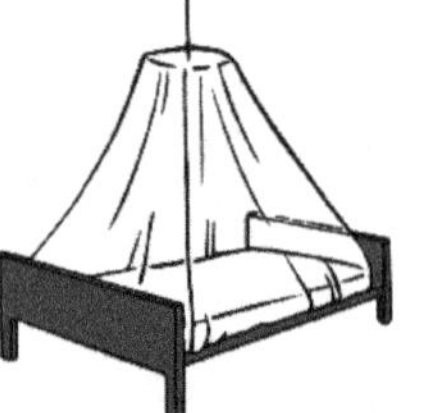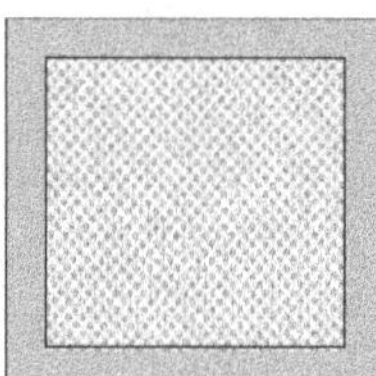

Fig. 15.10: Nets over the bed and on windows to prevent being bitten

- Wear long sleeved clothes to prevent being bitten by mosquitos.
- Use insect repellent on the skin to avoid being bitten by mosquitos.

Discourage mosquitos
- Use insecticides to kill adult mosquitos.
- Paint the wall white and wear light coloured clothes.

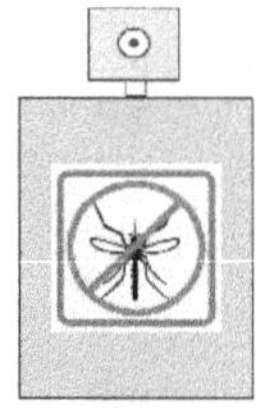

Fig. 15.11: Mosquito repellents or insecticides

Uses of tropical rain forests

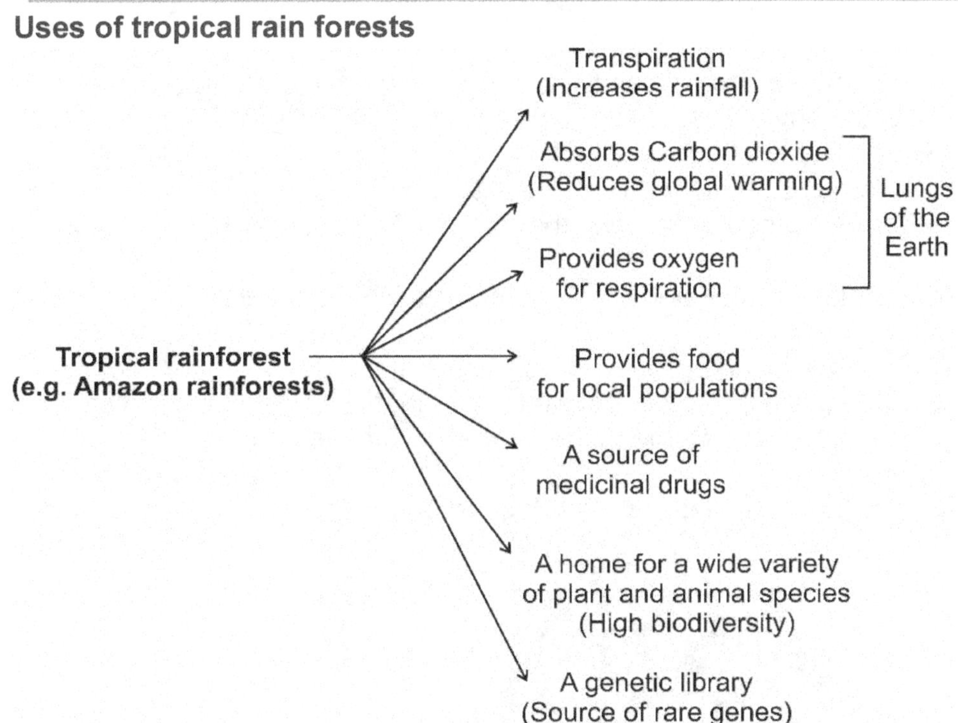

Effects of humans on tropical rainforests

1. Deforestation for timber and urbanisation reduces the area covered by forests.

2. Acid rain due to pollutants from burning of fossil fuels causes death of plants.

3. Loss of biodiversity and extinction of species due to poaching and habitat loss.

4. Building of roads through the forests leads to fragmentation of the forest and affects biodiversity.

Uses of oceans

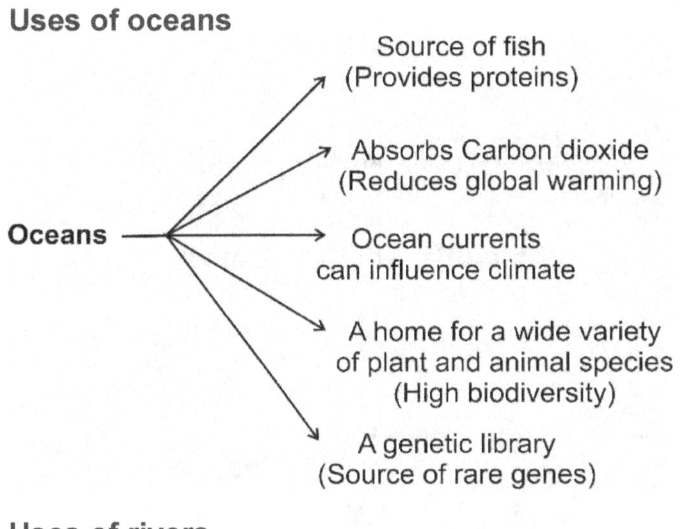

Effects of humans on oceans

1. Acidification of oceans due to acid rain can lead to death of algae, resulting in destruction of food chains and subsequent loss of biodiversity.

2. Global warming due to burning of fossil fuels can reduce carbon dioxide dissolution in the ocean. This can further accelerate global warming due to increase in carbon dioxide in the atmosphere.

3. Overfishing can lead to dwindling of fish stocks.

Uses of rivers

Effects of humans on rivers

1. Disposal of untreated sewage results in eutrophication, algal blooms and loss of biodiversity.

2. Soil erosion due to deforestation can result in silting and flooding of rivers.

3. Pesticides from agricultural land can drain into rivers and accumulate in the food chains. This can result in fish with high concentration of pesticides in their bodies.

Cambridge 5090 syllabus specification 15(k) describe the consequences of deforestation in terms of its effects on soil stability, climate and local human populations;

Effects of deforestation

- The roots of plants bind soil. So, deforestation increases soil erosion as loss of roots loosens the soil. Forests also reduce wind speed and the canopy prevents rain beating down on soil. The eroded soil is carried into rivers and causes flooding.
- Deforestation increases the rate of surface run off, causing rain water to run into streams or rivers rather than being held in soil. This increases loss of nutrients from the soil through leaching. This leads to eutrophication of fresh water bodies.
- Less carbon dioxide is removed from the atmosphere and more is added. This adds to the problem of global warming.
- Deforestation results in a loss or reduction of biodiversity due to loss of habitats and food sources of many organisms. Reduction in biodiversity leads to a limited gene pool with less variation. This increases chance of extinction of the species in a changing environment. Some of the species that have become extinct might have proven to be useful to us.
- Deforestation results in economic losses to local populations who depend on ecotourism to attract visitors to natural forests, as in Africa.

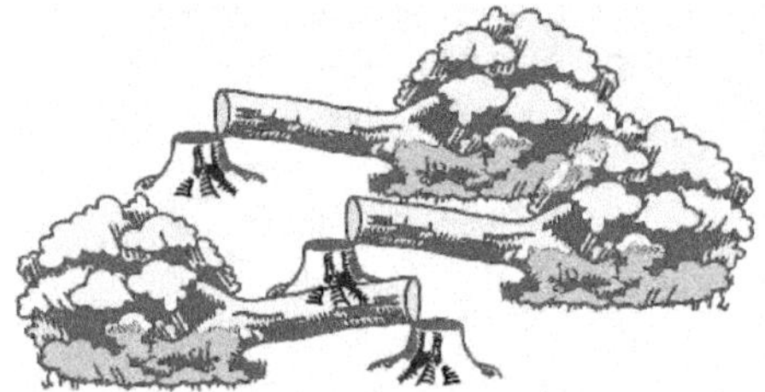

Fig. 15.12: Deforestation

Cambridge 5090 syllabus specification 15(l) evaluate the effects of: water pollution by sewage, by inorganic waste and by nitrogen-containing fertilisers; air pollution by greenhouse gases (carbon dioxide and methane), contributing to global warming; air pollution by acidic gases (sulfur dioxide and oxides of nitrogen), contributing to acid rain; pollution due to insecticides;

Water pollution by Sewage, inorganic waste (phosphates from detergents) and fertilisers

Figure 15.13 shows the effects of organic pollution in a river or lake.

Eutrophication
(Excess minerals in water)
↓
Algal blooms form due to rapid growth
↓
Light blocked and many algae die
↓
Dead algae are decomposed by bacteria, which use up oxygen rapidly in water
↓
Lack of oxygen kills fish and other aquatic life

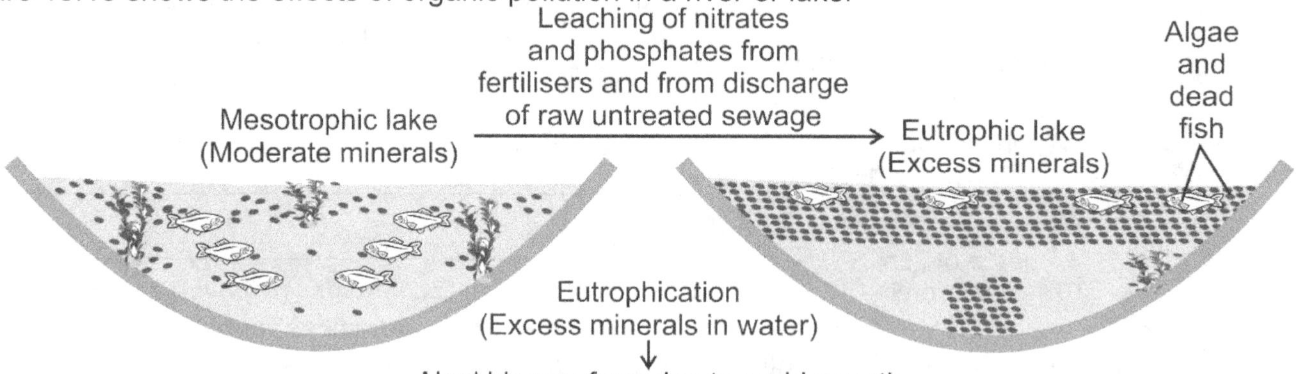

Fig. 15.13: Eutrophication of a river or lake

Air pollution by greenhouse gases (carbon dioxide and methane)

Human activities, like burning of fossil fuels, raring of cattle, deforestation and increase in landfill sites, in recent years have led to **an increase in release of greenhouse gases**. The greenhouse gases trap long wavelength infrared radiation and keep the atmosphere warm, as shown in figure 15.14. An increase in the greenhouse effect has led to an **increase in mean global temperatures** commonly referred to as **Global warming.**

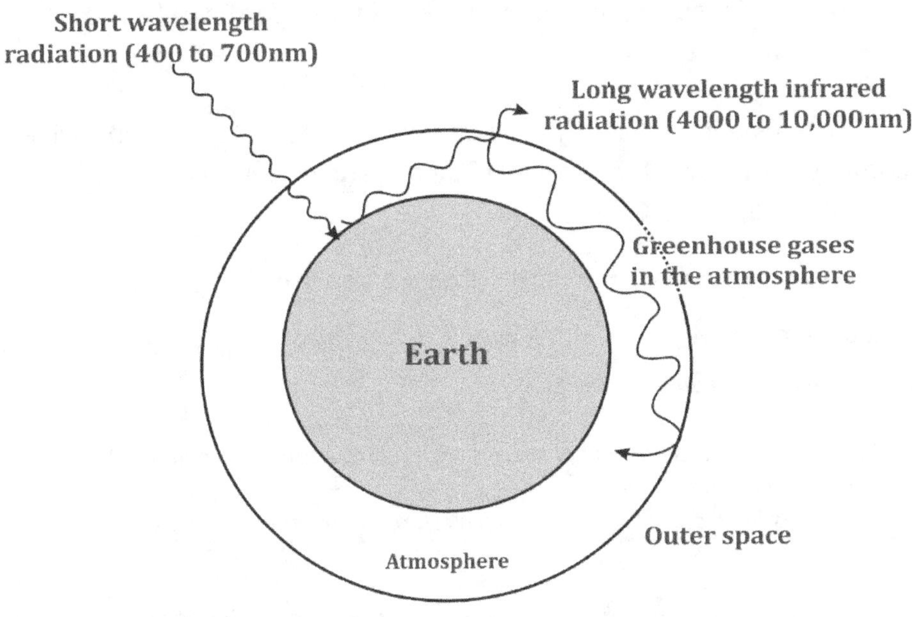

Fig. 15.14: Global warming due to increase in Greenhouse effect

Effects of global warming

1. Melting of polar ice caps and subsequent sea level rise.

2. Changes in the distribution of species. Species will move to cooler places and colonise new areas.

3. Change in the migration pattern of birds.

4. Climate change: erratic rainfall, changes in the seasons, warmer summers, etc.

5. Speeding up the life cycle of insects.

6. Increase in the rate of photosynthesis and decomposition.

7. Decrease in solubility of carbon dioxide in the ocean.

Air pollution by acidic gases (sulfur dioxide and oxides of nitrogen)

Burning fossil fuels releases nitrogen oxides and sulphur dioxide. These gases dissolve in rain water in clouds to form Nitric acid, Nitrous acid, Sulphurous and Sulphuric acid, as shown in figure 15.15. These acids are carried in clouds to long distances and the acid precipitation has harmful effects on plants, soil and animals.

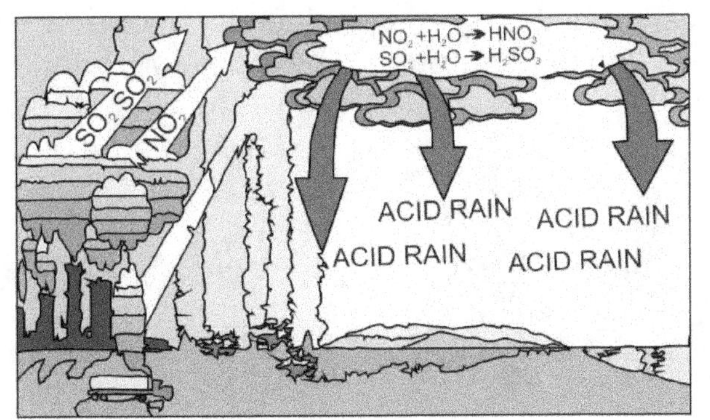

Fig. 15.15: Acid rain

Effects of acid rain

1. Acid rain decreases pH of soil. Microorganism activity is reduced and recycling of nutrients decreases. This reduces nutrient content and fertility of the soil. Plant growth decreases.

2. Acid rain damages roots of plants, decreasing ion uptake. It also damages the cuticle of leaves causing death of leaves by excess transpiration, especially in conifers on mountain slopes. Conifers become weak and susceptible to disease and drought.

3. Acid rain causes leaching of Al^{3+} ions from soil into rivers and lakes. Al^{3+} ions cause excess secretion of mucous in gills of fishes, reducing gas and ions exchange, causing death of fishes.

4. Acid rain inhibits the enzyme which causes hatching of fish and amphibian eggs. Reducing fish and amphibian population.

5. Heavy metals in water cause diseases in humans. Mercury poisons the foetus. Copper causes hair to turn greenish.

Pollution due to insecticides

Some pesticides are **persistent** so do not break down in the environment or within the tissues of living organisms. This gives rise to two potential problems:

a) **Bioaccumulation** - the accumulation of a substance in living tissue. Organisms at any trophic level may be capable of bioaccumulation.

b) **Bio-magnification** is the increasing concentration of a substance up a food chain, i.e. from one trophic level to the next . Animals at the higher trophic levels will be most affected.

Example

Some of the earliest insecticides, such as DDT, persists in fatty tissues (**bioaccumulation**). Larger, long lived predators at the end of a food chain may accumulate a lethal quantity of DDT as a result of eating large numbers of smaller species (**biomagnification**), as shown in figure 15.16.

Food chain   Water $\longrightarrow$ Plankton $\longrightarrow$ Small fish $\longrightarrow$ Large fish $\longrightarrow$ Birds

Concentration of pesticide in tissues 0.001ppm   0.04 ppm   0.5 ppm   2 ppm   25 ppm

Fig. 15.16: Bio-magnification of pesticides

> *Cambridge 5090 syllabus specification 15(m) discuss reasons for conservation of species with reference to maintenance of biodiversity, management of fisheries and management of timber production;*
> *Cambridge 5090 syllabus specification 15(n) discuss reasons for recycling materials, with reference to named examples.*

Reasons for conservation of species

- Many species of plants provide antibiotics and other useful drugs. So, it is necessary to conserve plant species to ensure that the potential cures for many diseases are not lost forever.
- Every species has a specific role in the food chain and food webs. Conserving species helps to reduce damage to food chains and food webs.
- Many species of fish and plants form a major component of our diet. It is vital to protect and sustain our food supply. Fisheries management helps to prevent overfishing and sustain fish stocks. Using large holes in fish nets helps to catch the larger fish only and allows the smaller fish to grow and reproduce. This helps to maintain fish stocks.
- Forests are a major source of timber (wood). Destruction of forests for timber could result in permanent loss of many useful species and genes. Timber management by using wood from energy plantations provide a continuous supply of wood by *short – rotation coppicing*. Fast growing woody species, like willow and poplar, which are easy to propagate (grow from stem cuttings) and produce high yield, are grown for 2 to 3 years and then cut off (coppiced) 2 to 3 feet above the ground. The stools or stumps left behind in the ground produce many shoots or poles which can be harvested every 2 to 5 years, as shown in figure 15.17.

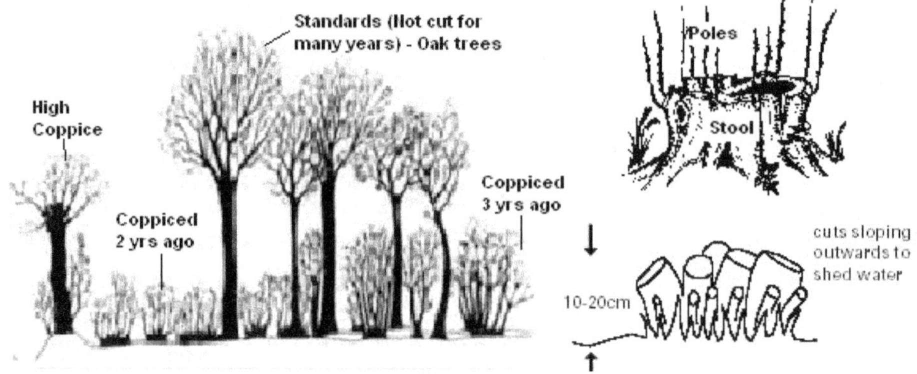

Fig. 15.17: A coppiced woodland

Recycling of materials

Recycling of paper helps to reduce deforestation, as paper is made from wood pulp.

Recycling of plastics is useful as plastics are non-biodegradable and do not decay. They persist in the environment and damage coral reefs and reduce drainage in soil.

> *Cambridge 5090 syllabus specification 16(a) define mitosis as cell division giving rise to genetically identical cells in which the chromosome number is maintained and state the role of mitosis in growth, repair of damaged tissues, replacement of worn out cells and asexual reproduction;*

Mitotic cell division

During mitosis, a single parent cell divides to form 2 identical daughter cells which have the same number of chromosomes as the parent cell, as shown in figure 16.1. The daughter cells are genetically identical to each other.

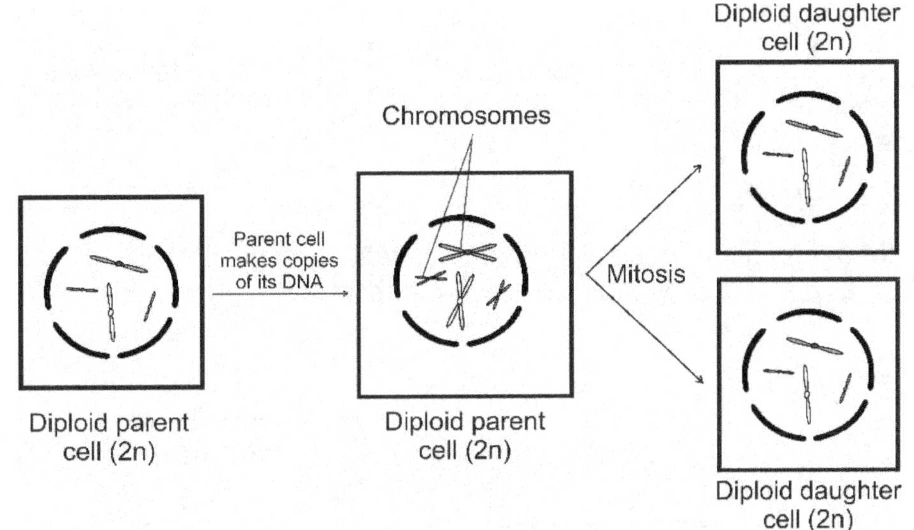

Fig. 16.1: Mitosis

- Mitosis is used to replace damaged cells and repair damaged tissues (wound healing).
- Mitosis is also needed for growth of organisms. Humans grow into huge organisms with trillions of cells, formed by mitosis, from a single celled zygote.
- Mitosis is used in asexual reproduction. A single plant can be cloned by using stem cuttings.

> *Cambridge 5090 syllabus specification 16(b) define asexual reproduction as the process resulting in the production of genetically identical offspring from one parent and describe one named, commercially important application of asexual reproduction in plants;*

Asexual reproduction results in the formation of genetically identical offspring from a single parent. It does not involve meiosis or the formation of gametes. The commercial benefits of asexual reproduction in plants are listed below.

- Mutation may give rise to a rare plant with useful characteristics.
- For example a plant which produces very high yield of rice, or a plant that is able to resist drought (lack of water), or a plant which is resistant to attack from pests.
- If these plants are allowed to reproduce sexually, then the characteristics may not be passed on to the offspring.
- Asexual reproduction can be used to produce a large number of genetically identical plants in a very short period of time. All the offspring will have the exact same features of the parent plant.

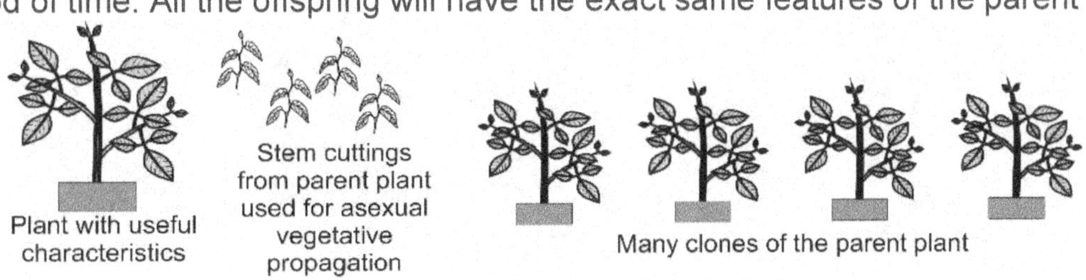

Fig. 16.2: Asexual reproduction (cloning)

Meiosis

A Diploid cell or organism has chromosomes found in pairs. There are two sets of chromosomes. Somatic cells (body cells) are diploid. A diploid cell from human skin will have 23 pairs of chromosomes (46 chromosomes).	In a **haploid** cell the chromosomes are not found in pairs. There is only one set of chromosomes. Sex cells (egg and sperm cells) contain a single set of chromosomes and are known as haploid. A human sperm cell will have only 23 chromosomes.

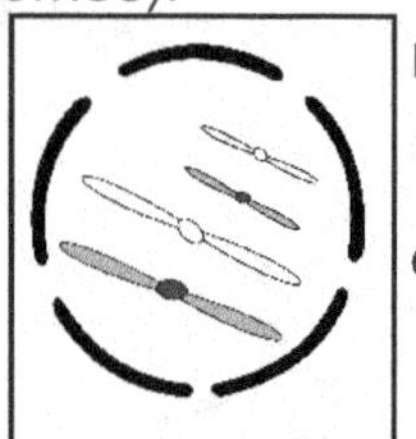

Fig. 16.3: Diploid cell

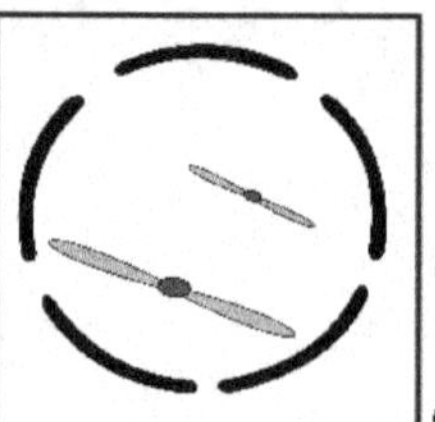
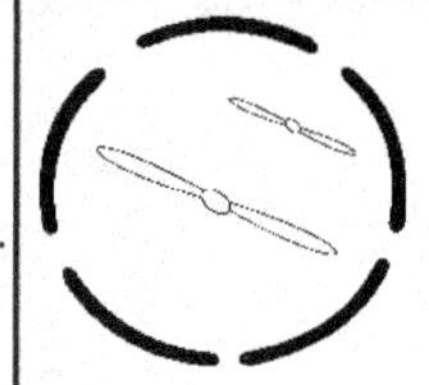

Fig. 16.4: Haploid cell

Meiosis results in the formation of 4 haploid daughter cells from a single diploid parent cells. The daughter cells form gametes. All the gametes are genetically non-identical.

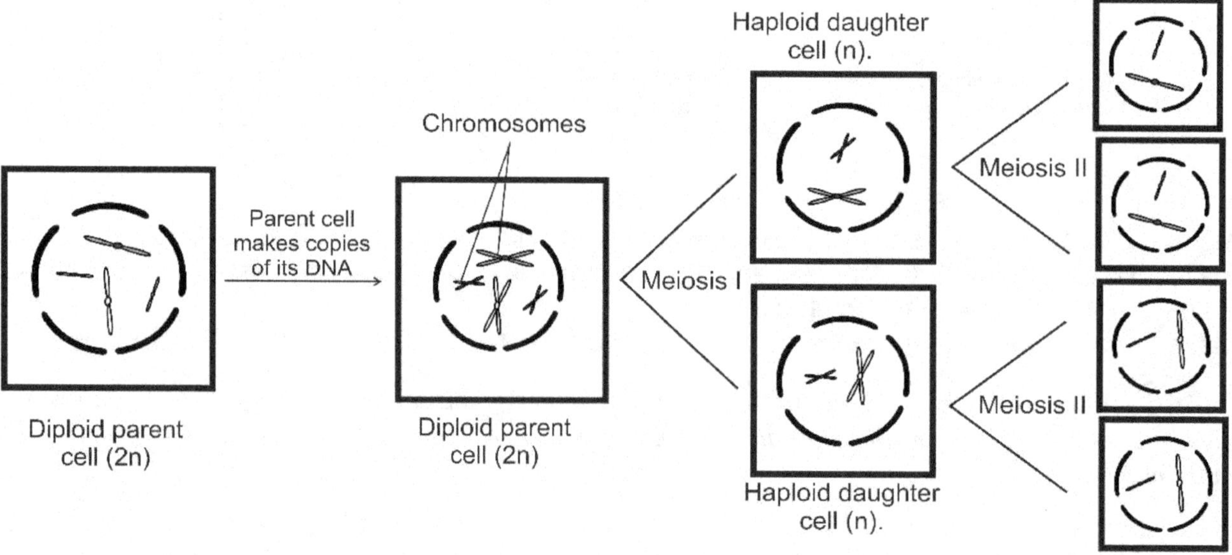

Fig. 16.5: Meiosis

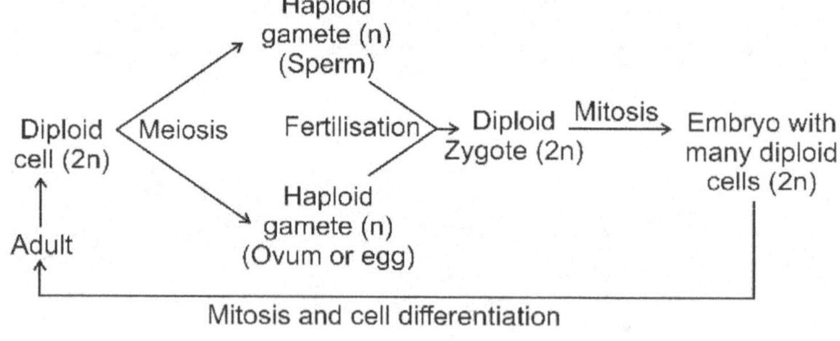

Fig. 16.6: Sexual reproduction

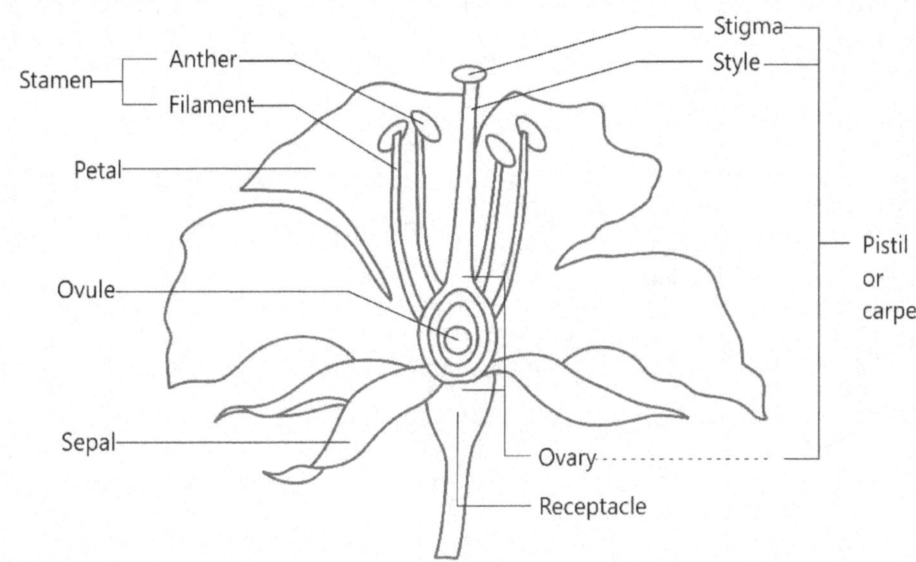

Fig. 16.7: A typical insect pollinated dicotyledonous flower

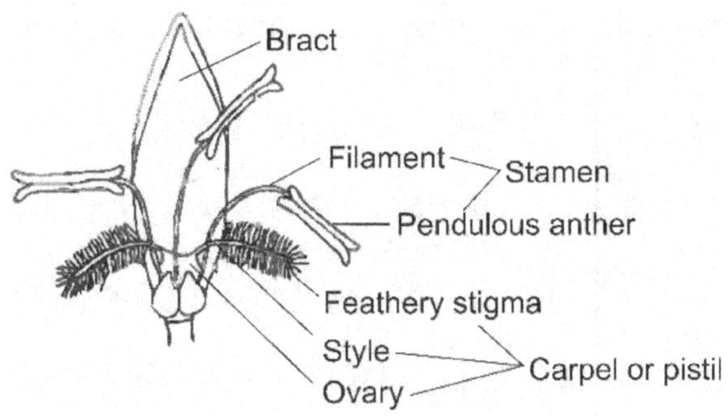

Fig. 16.7: A typical wind pollinated flower

Figure 16.8 shows the appearance of pollen from plants of different species. The pollen are also specially adapted to their mode of pollination.

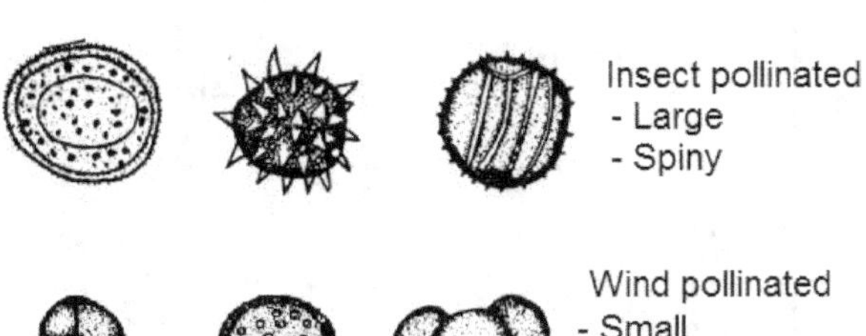

Fig. 16.8: Pollen structure

Anthers

Anthers produce the pollen grains, which contain the male gametes. Filaments hold the anther in place to facilitate pollination. In wind pollinated flowers the filament is long and the anthers hang out of the flower (pendulous).

Sepals

Sepals are usually green in colour and can photosynthesise. The sepals also protect the flower during the bud stage.

Petals

Petals are usually large and brightly coloured in insect pollinated flowers. They provide a landing platform for insects which are attracted to the bright colour and sweet smell of flowers. This helps in pollination. The petals have a nectary at the base, which produces nectar to attract pollinators.

Carpel

The carpel consists of three regions:

- **Stigma**: receives the pollen during pollination. It is sticky in insect pollinated flowers and feathery in wind pollinated flowers.
- **Style**: is a tubular structure which connects the stigma to the ovary. It holds the stigma in a favourable position for pollination.
- **Ovary**: the ovary holds the ovules. The ovary ripens to become the fruit after pollination is complete. The ovule contains the female gamete (ovum or egg cell).

- Pollination is the transfer of pollen from the anther to the stigma of a flower.
- If pollen is transferred from the anther to the stigma of a flower on the same plant then it is called as **self-pollination**, as shown in figure 16.9.
- If pollen is transferred from the anther to the stigma of a flower on another plant then it is called as **cross-pollination**, as shown in figure 16.9.
- Insects or wind may act as agents of pollination.

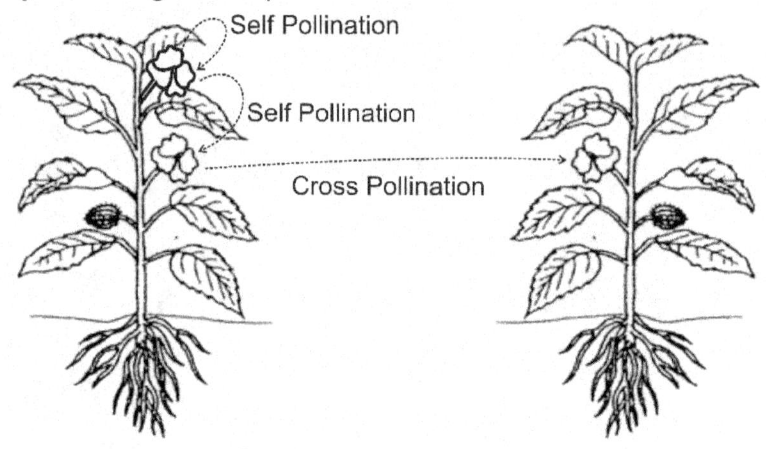

Fig. 16.9: Types of pollination

Insect pollinated flower	Wind pollinated flower
1. Large, brightly coloured petals to attract insects and act as a platform.	1. Petals are small, if present. Not brightly coloured.
2. Nectaries produce sweet, scented nectar, which insects feed upon.	2. Nectaries absent.
3. Stamens enclosed within the flower, usually in a position to transfer pollen onto the body of a visiting insect.	3. Stamens hang out of the flower (pendulous) to disperse pollen into the wind. Anthers are usually large to produce large quantities of pollen.
4. Larger pollen grains, with sculptured walls to help it attach to the body of the insect. Fewer pollen produced as chances of pollination are more.	4. Small, light, smooth pollen grains to be easily carried by wind. Large number of pollen are produced to increase chances of pollination.
5. Stigma is small and enclosed within the flower. Positioned in such a way that it comes into contact with the visiting insect, to pick up pollen.	5. Long, feathery, branched stigma hanging out of the flower to trap pollen from the wind.

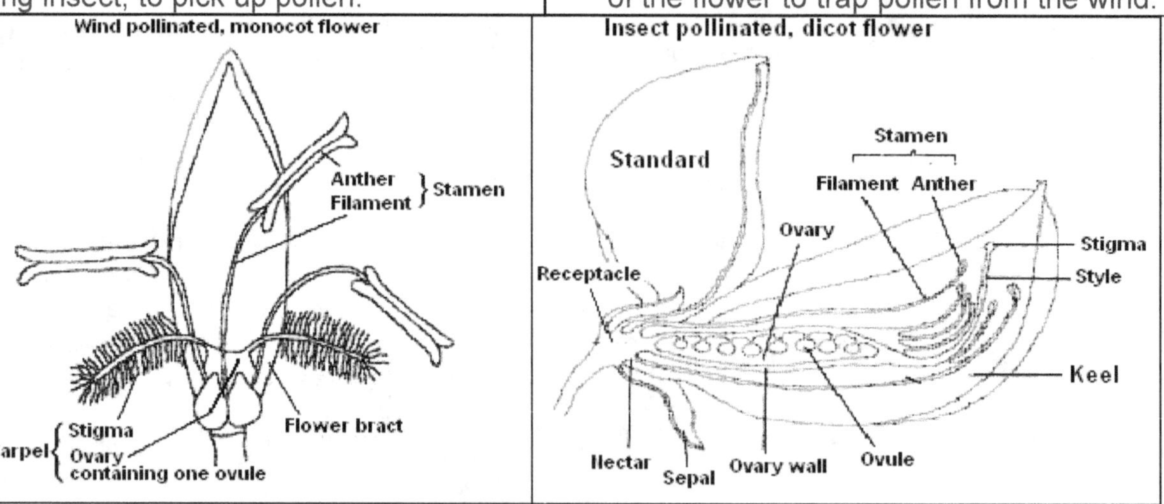

Fig. 16.10: Insect and wind pollinated flowers compared

After pollination the epidermal cells of the stigma secrete sucrose, with traces of boron. This stimulates the growth of the pollen tube. As the pollen tube grows, the Generative haploid nucleus splits by mitosis to form two haploid male nuclei. The pollen tube penetrates the tissues of the style to reach the ovule, through the micropyle.

The pollen tube has two main functions
1. It delivers the haploid (n) male nucleus (gamete) from the pollen grain, into the embryo sac - where the haploid (n) female gamete is present. This enables fertilization to occur.
2. The Pollen tube secretes auxins which stimulates the development of the ovary into a fruit.

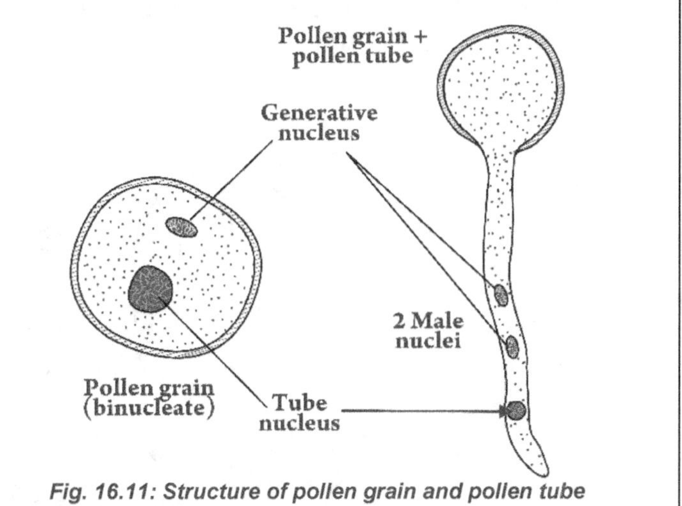

Fig. 16.11: Structure of pollen grain and pollen tube

Fertilisation

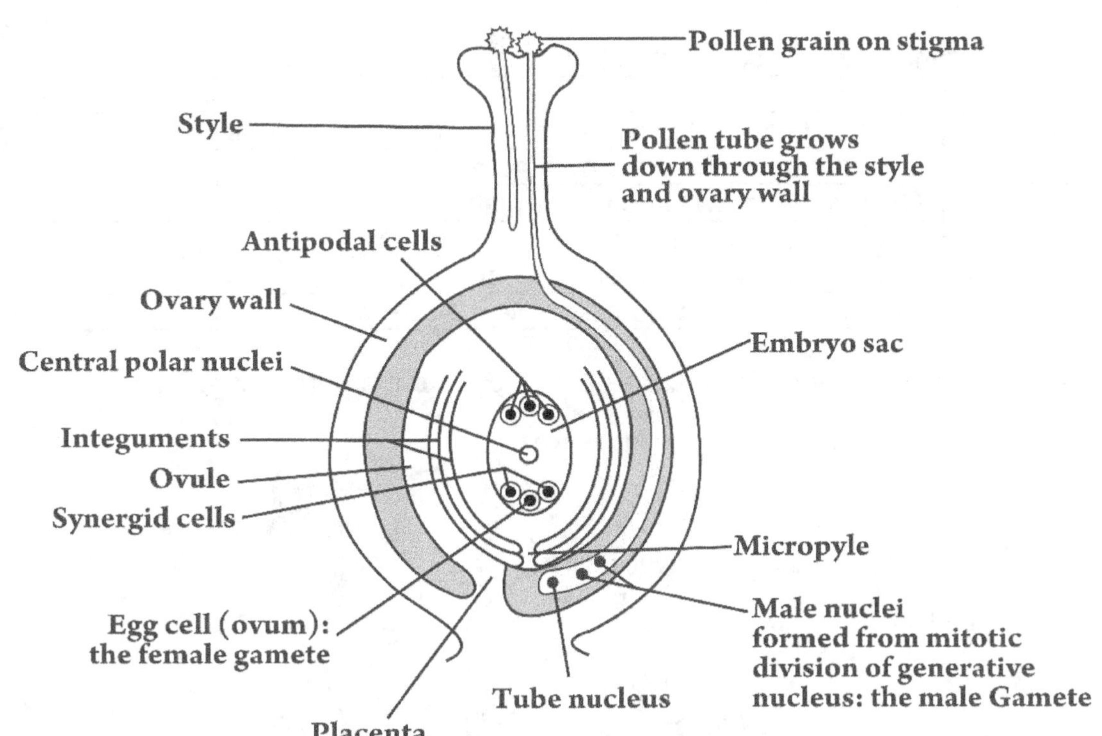

Fig. 16.12: Fertilisation in flowering plants

- The growth of the pollen tube into the embryo sac is controlled by enzymes, produced by the tube nucleus. These enzymes soften the middle-lamella between the cells of the style. The pollen tube is guided towards the micropyle by chemical signals.
- The pollen tube contains three nuclei – one tube nucleus and two haploid male nuclei.
- When the pollen tube enters the embryo sac through the micropyle, the tube nucleus degenerates. One haploid male nucleus (n) fuses with the haploid egg cell nucleus (n) to form a diploid zygote (2n). The zygote develops into the embryo of the seed.
- The other haploid male nucleus (n) combines with the two haploid polar nuclei (n + n) to form a triploid (3n) endosperm nucleus. The endosperm nucleus will develop into the endosperm of the seed, which stores food. This is called as **double fertilization,** because two male nuclei fuse with two female nuclei.

The fruit wall is called the pericarp. Seeds are found inside the pericarp. This provides protection to the seed and also helps in dispersal.

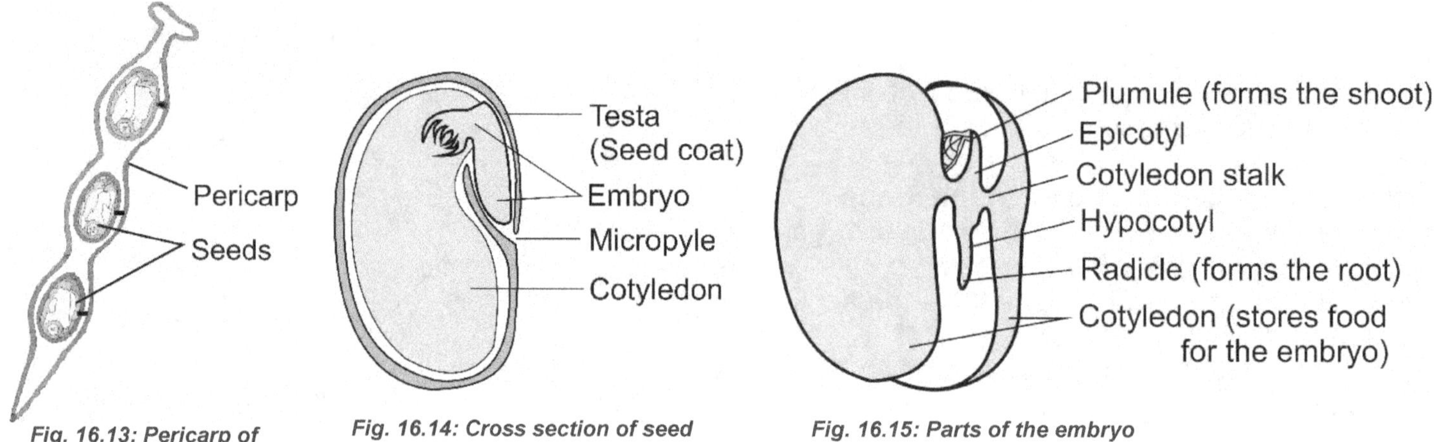

Fig. 16.13: Pericarp of bean

Fig. 16.14: Cross section of seed

Fig. 16.15: Parts of the embryo

The functions of the different parts of a seed are listed below.

Structure	Function
Testa or seed coat	The seed coat helps protect the embryo from mechanical injury and drying out.
Cotyledons	Stores food for the growth of the embryo.
Embryo	The embryo grows into a new plant.
Radicle	The radicle forms the root of the new plant.
Plumule	The plumule forms the shoot or stem.
Micropyle	Allows oxygen and water to enter the seed.

Figure 16.16 shows the parts of a seedling that has developed from a germinating seed.

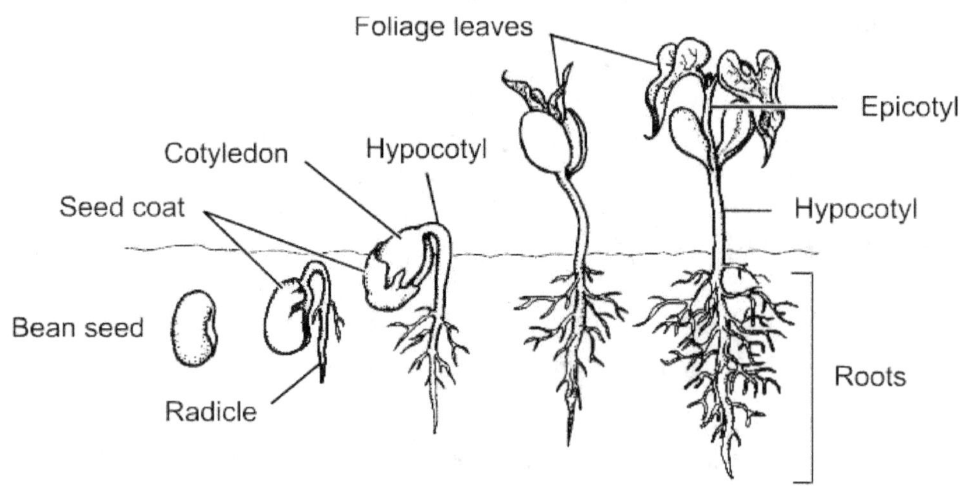

Fig. 16.16: Germination of a dicot seed

Seed dispersal

Seed dispersal is essential to **reduce competition** between the parent plant and the seedlings. It also helps plants to colonise new areas. The seeds may be dispersed by wind or animals.

Wind dispersed fruits and seeds have the following features:
- They have hairy extensions or wing-like extensions of the pericarp or testa to catch the wind and be carried far away from the parent plant.
- They are usually dry, small and light to enable them to be carried over long distances by the wind.

e.g. dandelion, sycamore, maple, drumstick.

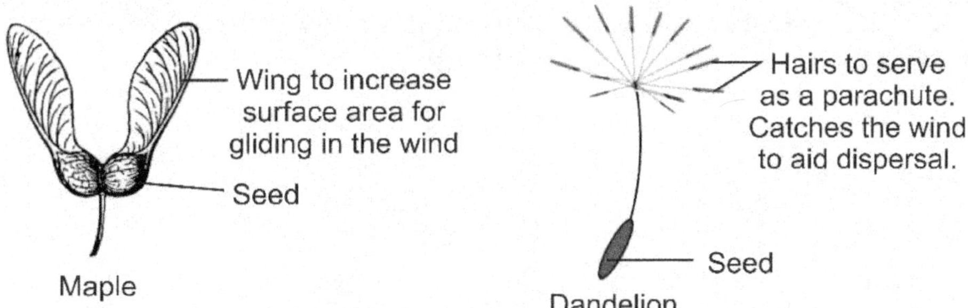

Fig. 16.17: Wind dispersed fruits

Animal dispersed fruits and seeds have the following features:
- They may be brightly coloured and succulent fruits which contain seeds with indigestible coats to allow the seeds to pass through the gut undamaged. e.g. tomato, plum, raspberry, grape, mango.
- They may have hooks on the pericarp which attach them to the fur of passing animals. e.g. goose grass, burdock, xanthium.

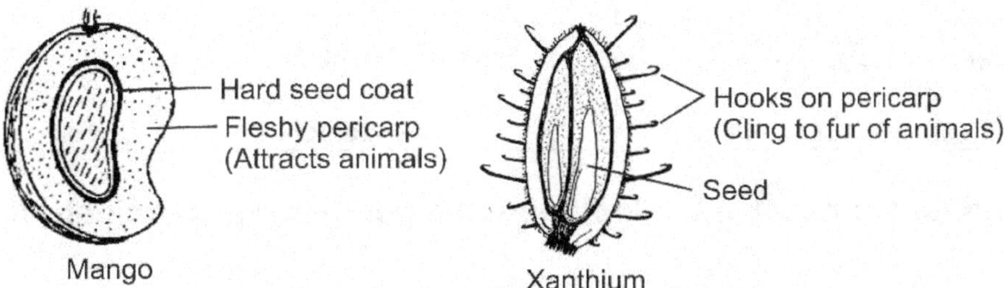

Fig. 16.18: Animal dispersed fruits

> *Cambridge 5090 syllabus specification 16(o) investigate and state the environmental conditions that affect germination of seeds: suitable temperature, water and oxygen;*
> *Cambridge 5090 syllabus specification 16(p) describe the uses of enzymes in the germination of seeds;*

Water, oxygen and an optimum temperature are necessary for a seed to germinate.
Water is absorbed through the micropyle and the testa. Water activates and dissolves the enzymes substrates in the cotyledons. The cotyledons swell and the testa splits open.

The enzymes in the cotyledon are used to hydrolyse the stored food into soluble compounds which can be used to provide energy and nutrients for the growth of the embryo. The enzymes involved in the hydrolysis of food in the cotyledons are shown in figure 16.19. Figure **16.20** shows changes **before the leaves** are formed.

Starch $\xrightarrow{\text{Amylase}}$ Maltose $\xrightarrow{\text{Maltase}}$ Glucose

Proteins $\xrightarrow{\text{Protease}}$ Amino acids

Lipids $\xrightarrow{\text{Lipase}}$ Fatty acids and glycerol

Fig. 16.19: Enzyme action in cotyledons of germinating seeds

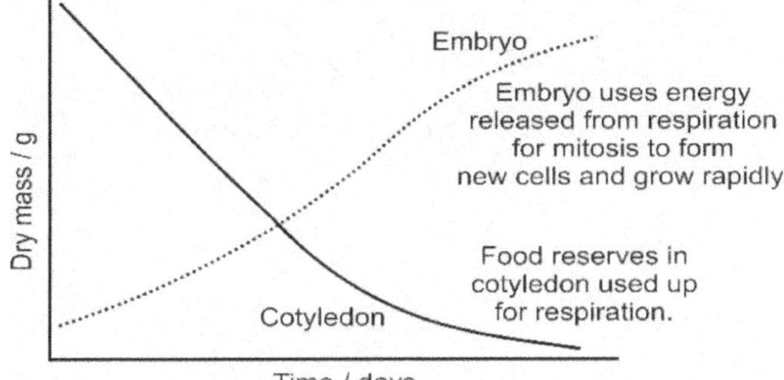

Fig. 16.20: Changes in the mass of embryo and cotyledons during germination

Oxygen is needed for respiration. The glucose is used as a respiratory substrate to provide energy for mitosis and active transport in the seedling.

An **optimum temperature** is required for the enzymes and substrates in the seed to have enough kinetic energy to react rapidly.

The experiments described below can be used to investigate the necessity of water, oxygen and optimum temperature for the germination of seeds.

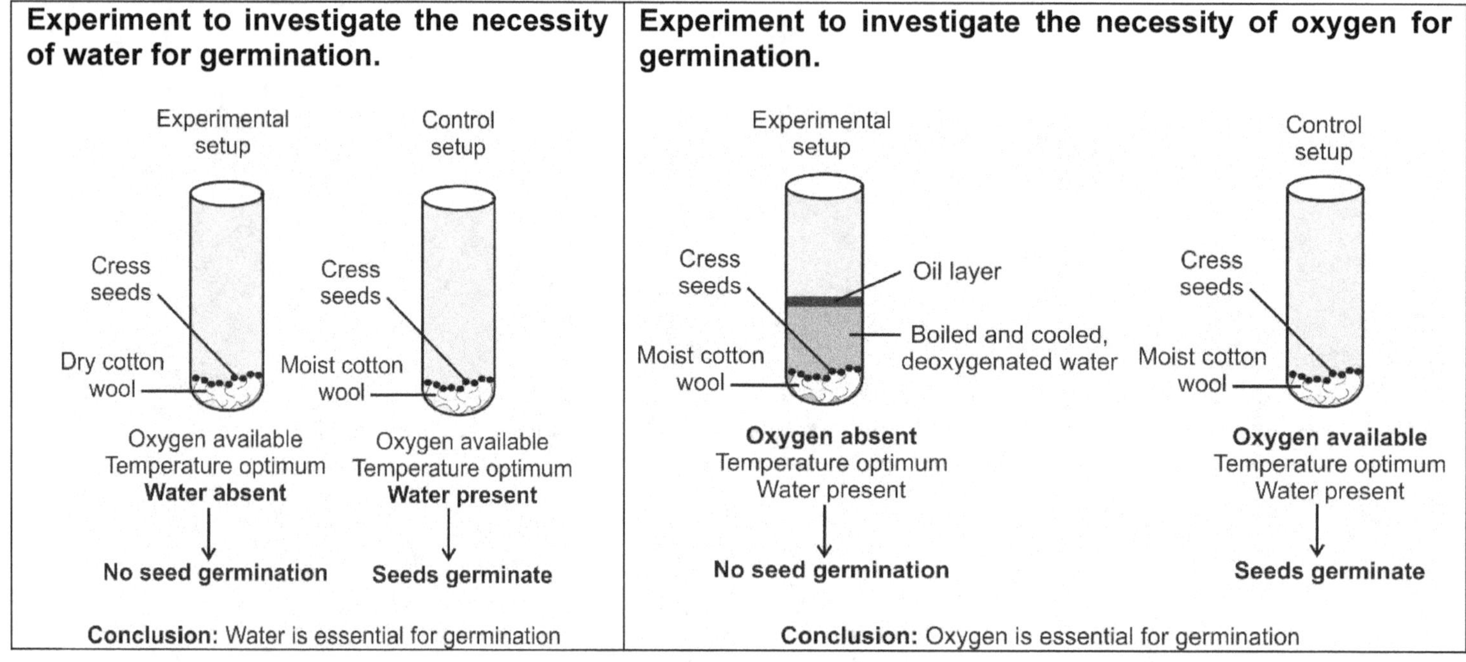

Experiment to investigate the necessity of optimum temperature for germination.

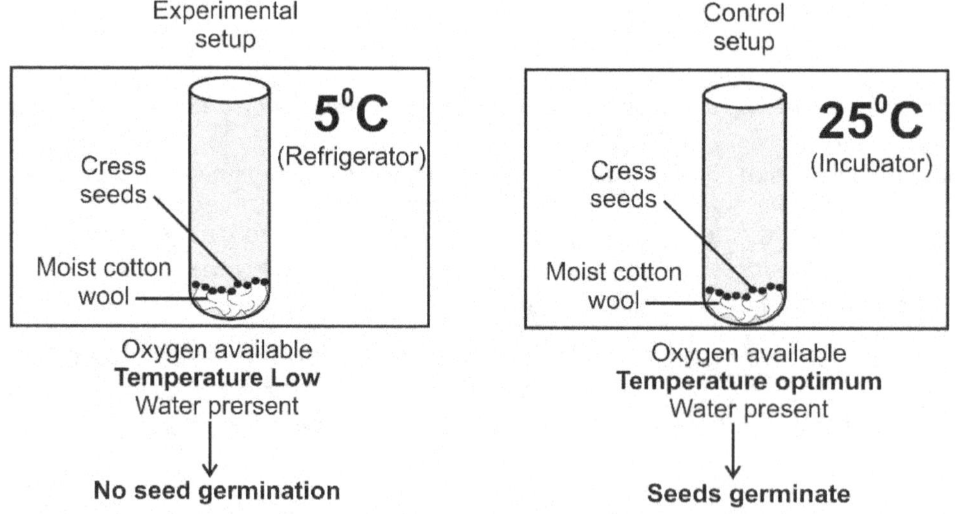

Conclusion: An optimum temperature is essential for germination

Male reproductive system

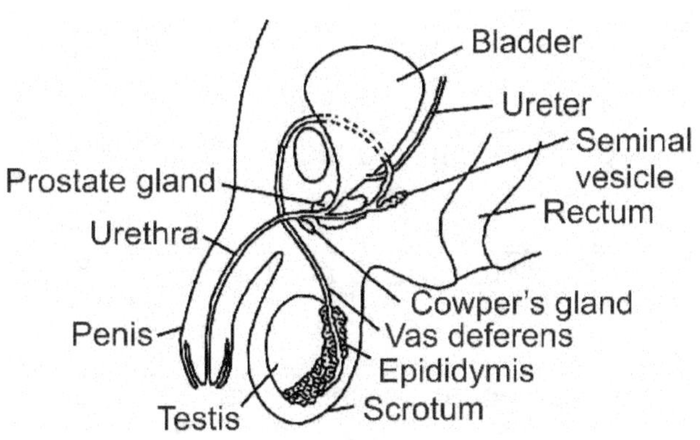

Fig. 16.21: Male reproductive system

- **Testis:** produces sperm and male sex hormone – testosterone.
- **Scrotum:** regulates the temperature of testis and maintains it 1°C lower than body temperature.
- **Epididymis:** stores sperms.
- **Vas deferens:** transfers sperms from epididymis to urethra by peristalsis.
- **Cowper's gland, prostate, seminal vesicle:** produces seminal fluid for lubrication during copulation, provides an alkaline medium for sperms to swim and obtain nourishment.
- **Penis**: transfers sperms to vagina by copulation.
- **Urethra**: common passage for sperm and urine.

Female reproductive system

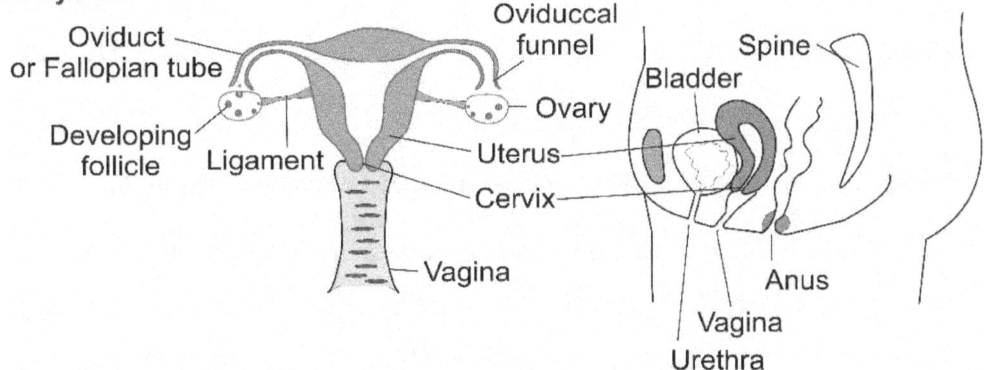

Fig. 16.22: Female reproductive system

Ovaries: Produce ovum and female sex hormones – oestrogen and progesterone.

Oviduct or fallopian tube: carries the ovum towards the sperms by cilia and peristalsis. Fertilisation takes place in the oviduct.

Uterus: a muscular organ which harbours and nourishes the growing embryo during pregnancy. The contraction of uterus muscles helps to push the baby out of the birth canal during parturition (child birth).

Cervix: the sperms can enter the uterus through the cervix.

Vagina: a muscular organ which receives the penis during copulation. It receives the sperms during copulation.

Sperm (male gamete)	Ovum (female gamete)
Smaller than the ovum and is motile (it can move).	Large cell, incapable of independent movement.
Long tail for swimming, powered by energy from within the cell.	Wafted along oviducts by cilia and muscular contractions of the oviduct.
Has little or no food reserves. It gets energy from nutrients in the seminal fluid.	Cytoplasm contains protein and lipid food reserves.
About 300 million sperms are released in a single ejaculation. This increases the chances of fertilisation as many sperms can release a large quantity of acrosomal enzymes to digest the jelly coat around the ovum.	A single ovum is produced and released during ovulation, once in 28 days.

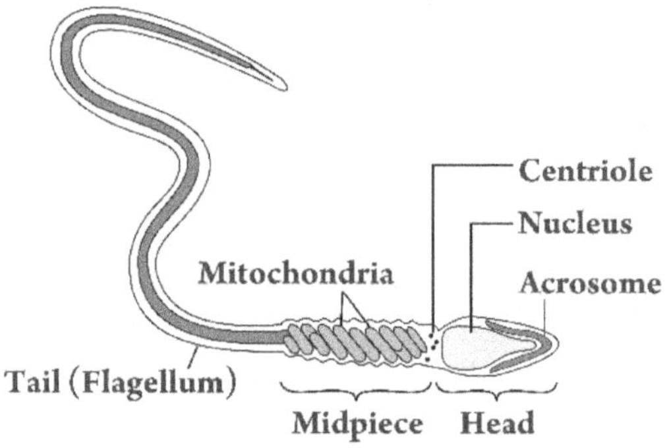

Centriole
Nucleus
Mitochondria
Acrosome
Tail (Flagellum)
Midpiece   Head

Fig. 16.23: Structure of sperm

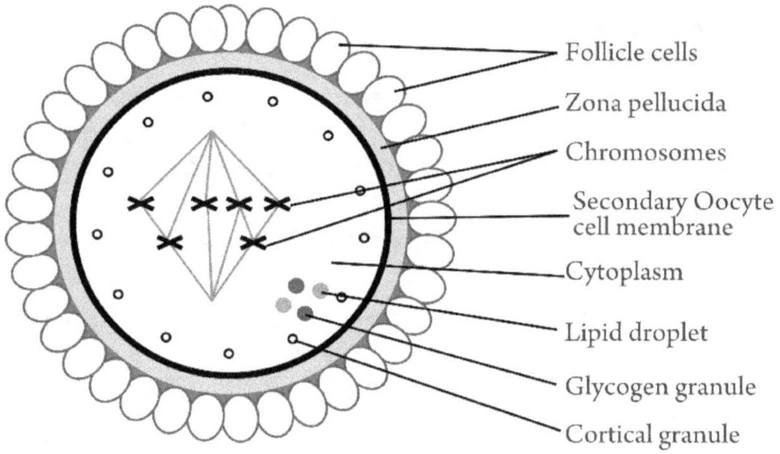

Follicle cells
Zona pellucida
Chromosomes
Secondary Oocyte cell membrane
Cytoplasm
Lipid droplet
Glycogen granule
Cortical granule

Fig. 16.24: Structure of ovum

Menstrual cycle

The average length of the menstrual cycle is 28 days. However there may be variation due to pregnancy or breast-feeding, dietary disorders or natural variations in different individuals.

The main stages are

Menstrual phase: days 1 to 5. This is accompanied by the shedding of the uterus wall.

Follicular phase: days 6 to 11. The follicles in the ovaries start to develop into mature Graafian follicles.

Ovulatory phase: the ovum is released from the follicle. It usually occurs on day 14.

Luteal phase: days 17 to 28. The corpus luteum forms from remnants of the follicle. It secretes progesterone which maintains the thickness of the uterus wall in preparation for implantation.

The menstrual cycle is controlled by the hormones Follicle Stimulating Hormone (FSH) and Luteinising Hormone (LH) produced by the pituitary gland and progesterone and oestrogen produced by the ovaries. Figure 16.25 shows the sequence of hormonal changes that regulate the menstrual cycle.

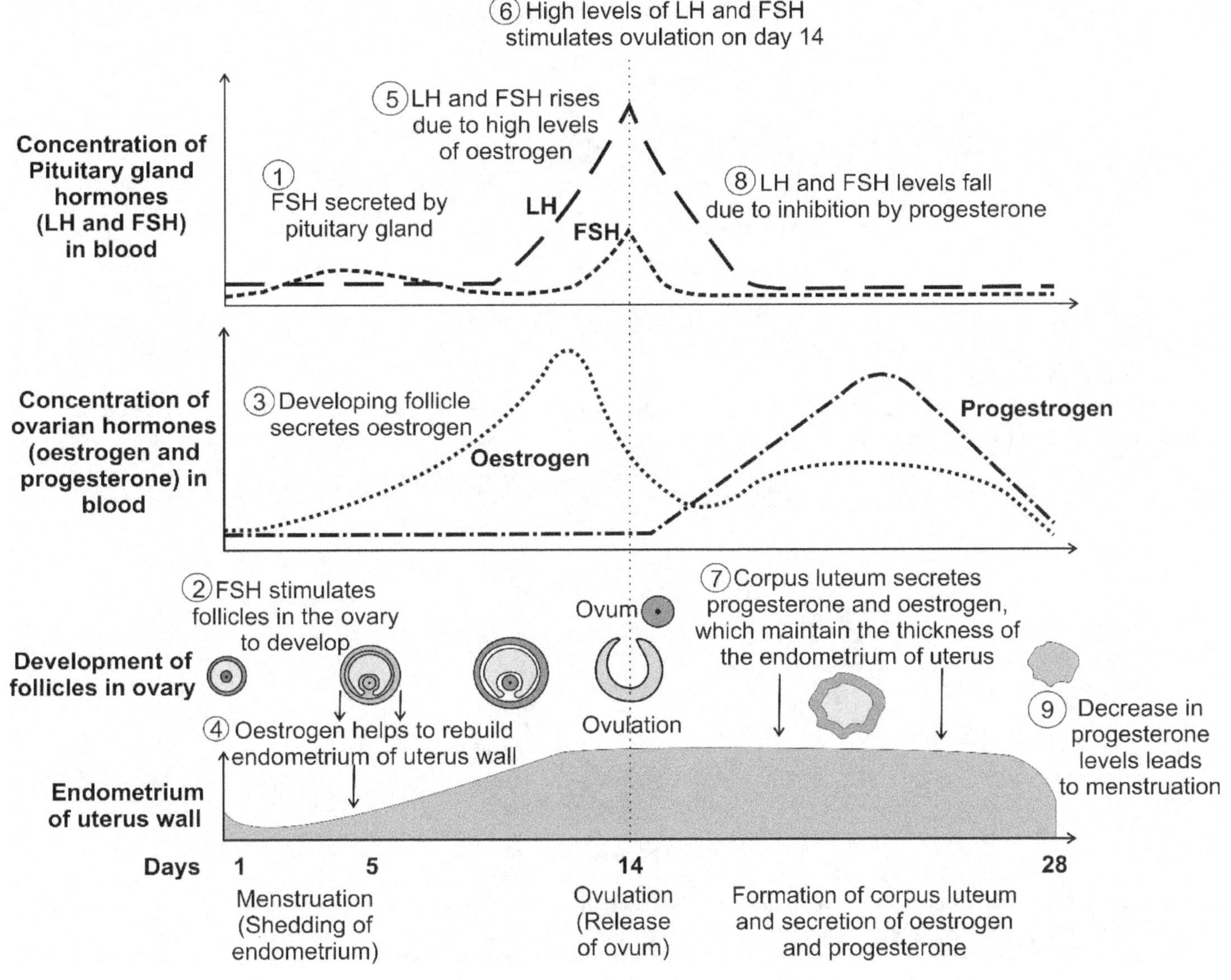

Fig. 16.25: Hormonal control of the menstrual cycle

Cambridge 5090 syllabus specification 16(v) describe fertilisation and early development of the zygote simply in terms of the formation of a ball of cells that becomes implanted in the wall of the uterus;
Cambridge 5090 syllabus specification 16(w) state the function of the amniotic sac and the amniotic fluid;

Fertilisation

- The semen is deposited at the cervix during copulation. The sperms then use their flagella to swim through the cervix and up through the uterus to the oviducts. The alkaline semen provides an alkaline medium for the sperms to swim. Out of the few millions of sperms, only a few hundred reach the oviduct and come into contact with the ovum.
- The ovum is surrounded by follicle cells and a clear membrane called the zona pellucida. Enzymes from the Acrosomes of many sperms hydrolyse a pathway through the follicle cells and zone pellucida.
- Eventually one sperm succeeds in passing through the outer layer and fuses with the cell surface membrane of the ovum. Immediately the cortical granules (lysosomes) rupture and release enzymes into the secondary oocyte, which causes the zona pellucida to thicken and prevent entry of other sperms.
- The sperm nucleus then enters the ovum. The male nucleus fuses with the female nucleus to form a diploid **zygote**. This is called **fertilisation**.

- The zygote then divides and re-divides by mitosis to form a ball of cells called a blastocyst. The outer layer of cells of the blastocyst is called the trophoblast, which is able to penetrate and get embedded into the uterus wall, as shown in figure 16.26. This is called **implantation**.

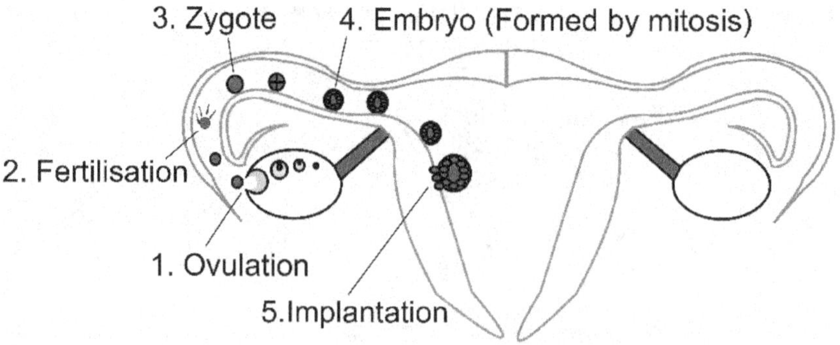

Fig. 16.26: Fertilisation and implantation

- The trophoblast develops into two membranes: the chorion and amnion. The amnion holds the amniotic fluid, as shown in figure 16.27.
- The amniotic fluid helps to absorb mechanical shock and injury. It also helps to keep the baby warm and maintains a constant temperature for the baby.

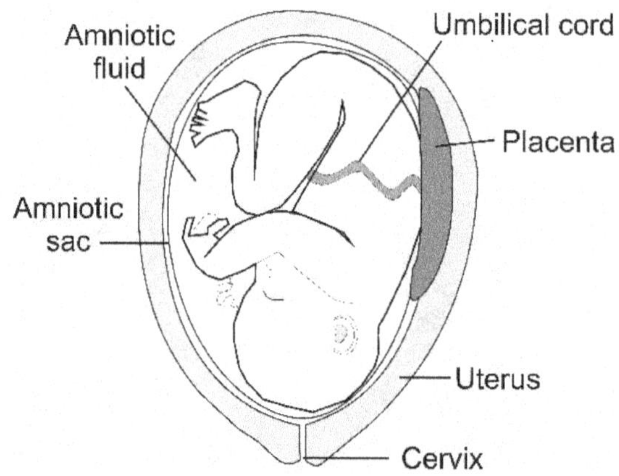

Fig. 16.27: Amniotic fluid

> *Cambridge 5090 syllabus specification 16(x) describe the function of the placenta and umbilical cord in relation to exchange of dissolved nutrients, gases and excretory products (no structural details are required);*

The placenta is an organ formed from the tissues of the mother and the foetus. The foetal capillaries are embedded into blood filled spaces within the placenta, as shown in figure 16.28. This increases the surface area for diffusion and prevents maternal and foetal blood from mixing.

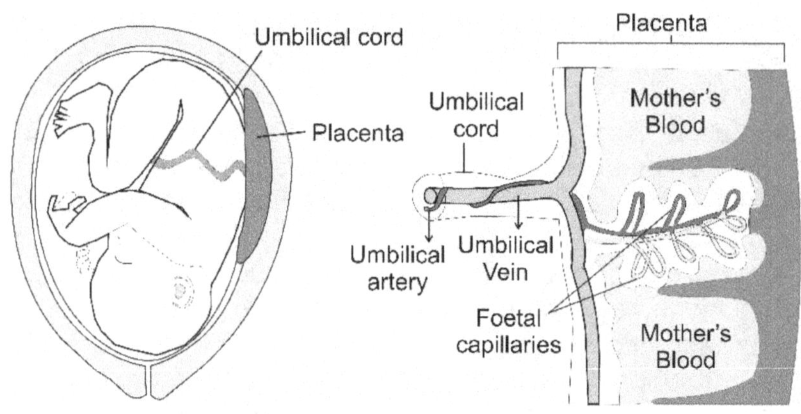

Fig. 16.28: Placenta and umbilical cord

The main functions of placenta include

- Diffusion of oxygen and food (Glucose, amino acids, vitamins and minerals) from maternal blood to foetal blood.
- Excretion of carbon dioxide and urea from foetal blood to maternal blood.
- The umbilical artery and umbilical vein then transport the blood between the mother and the foetus.
- Some antibodies of mother can cross the placenta, providing passive immunity to the foetus.

Pregnant women need a special diet as they have to provide nutrients for the growing foetus.

Iron: A diet rich in iron is essential for formation of foetal haemoglobin. The mother also needs a higher concentration of haemoglobin in the blood as the maternal blood must supply oxygen to the foetus as well.

Calcium and phosphate: A diet rich in calcium is essential for bone formation in the foetus.

Proteins: A diet rich in proteins will ensure rapid and healthy growth of the embryo and foetus.

Vitamins: A diet rich in vitamins is essential for the healthy development of the foetus.

Breast milk has the following advantages over bottle milk for a new born baby.
- Breast milk contains the correct proportions of dietary requirements for rapid growth and development of the baby.
- Breast milk contains antibodies from the mother. This helps to provide immunity to many pathogens.
- Breast milk is neither too warm nor too cold. It is at 37°C.
- Breast milk is inexpensive and readily available.
- The risk of infections is less as breast milk is sterile.

Methods of birth control
Natural method (Rhythm method)
The female is fertile only during a few days of the menstrual cycle. Avoiding sexual intercourse during the fertile period can help to prevent conception. Figure 16.29 shows the fertile period and the infertile periods of the menstrual cycle. The disadvantage is that the cycle is variable or irregular. This can lead to miscalculation and result in unwanted pregnancy.

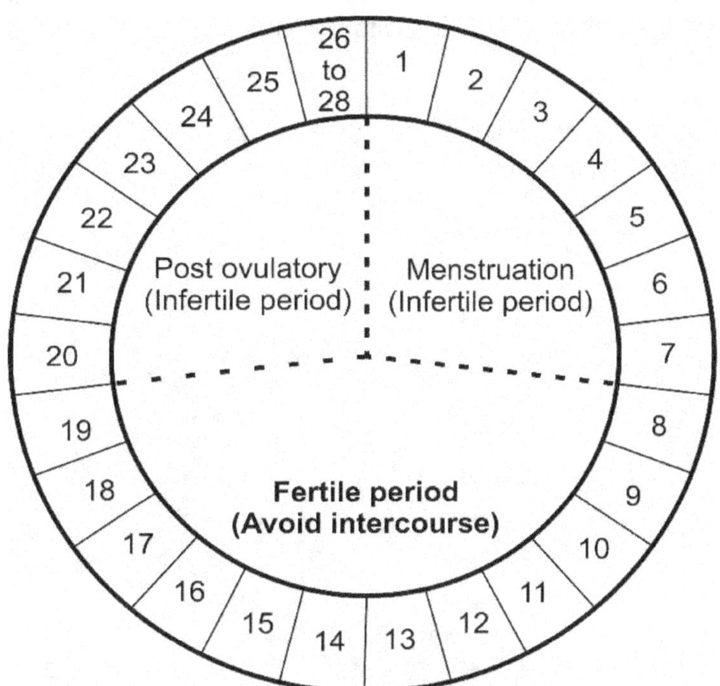

Fig. 16.29: Rhythm method of birth control

Natural method (Withdrawal method): The penis is withdrawn from the vagina just before ejaculation. However, the pre-ejaculant may contain some sperms and result in unwanted pregnancy.

Chemical method: spermicides are chemicals that kill sperms. The spermicides may be applied into the vagina before intercourse.

Mechanical method: the mechanical method prevents the sperms from meeting the ovum. In some cases it may prevent implantation of the embryo into the uterus.

Figure 16.30 summarises the mechanical methods of birth control.

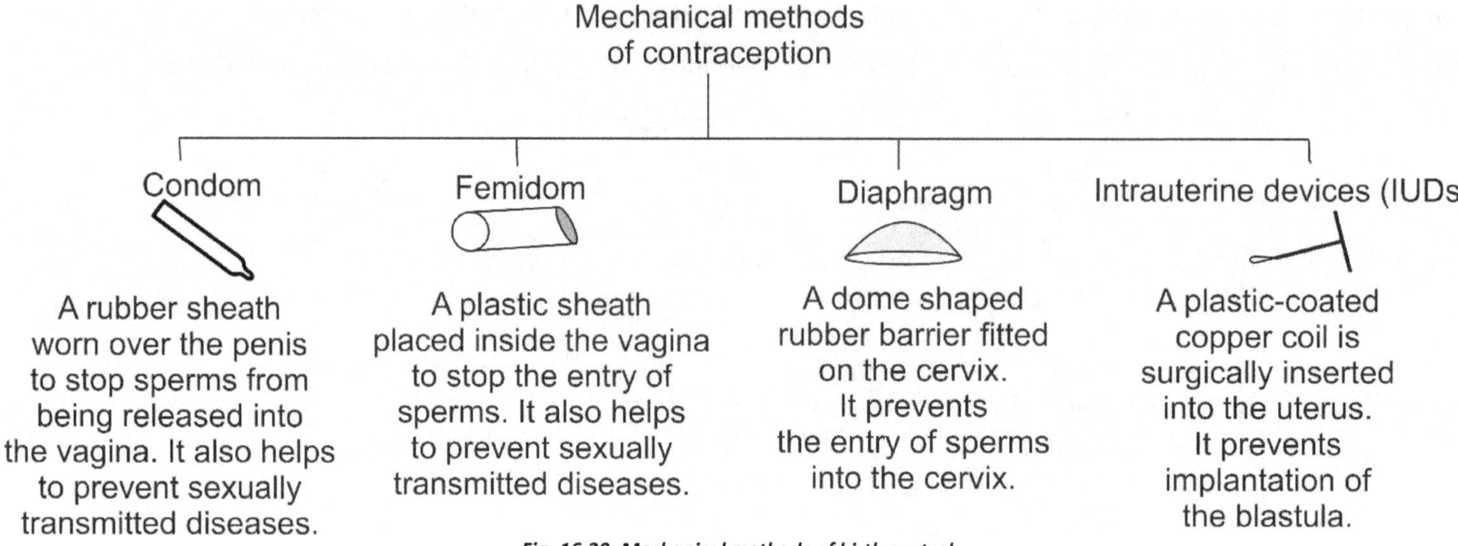

Fig. 16.30: Mechanical methods of birth control

Advantages of mechanical method of birth control	Disadvantages of mechanical method of birth control
<ul><li>Intrauterine devices are generally reliable.</li><li>Condoms are readily available.</li><li>Condoms protect from sexually transmitted diseases.</li></ul>	<ul><li>Intrauterine devices need to be fitted.</li><li>Condoms may split, resulting in pregnancy.</li><li>Condoms may result in loss of sensitivity during copulation.</li></ul>

Hormonal method: Progesterone inhibits the release of LH and FSH from the pituitary gland. Oral contraceptive pills contain progesterone. This inhibits the development of the follicles and prevents ovulation. So, pregnancy cannot occur.

Surgical method: The oviduct or sperm duct are cut and tied, as shown in figure 16.31. This prevents the sperms from coming in contact with the ovum.

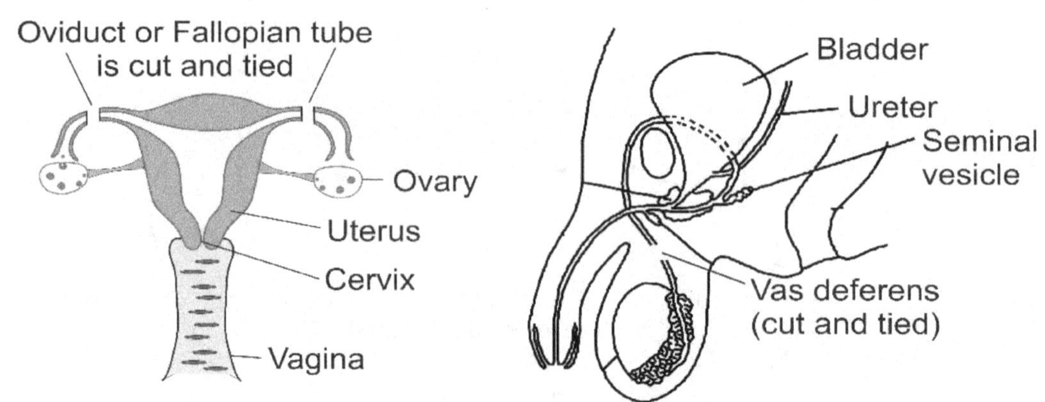

Fig. 16.31: Tubectomy and vasectomy

The advantage of surgical method is that it is very reliable. However, it is irreversible and permanent and is not advisable for young couples. Some people may also have a fear of surgery. In some cases the cost of surgery may also be high.

Cambridge 5090 syllabus specification 16(bb) explain that syphilis is caused by a bacterium that is transmitted during sexual intercourse;
Cambridge 5090 syllabus specification 16(cc) describe the symptoms, signs, effects and treatment of syphilis;

Syphilis is a sexually transmitted disease (STD) or venereal disease (VD). This means that it can be transmitted from an infected person to a non-infected person during sexual intercourse.
Syphilis is caused by *Treponema pallidum*, a spiral shaped bacterium, shown in figure 16.32.

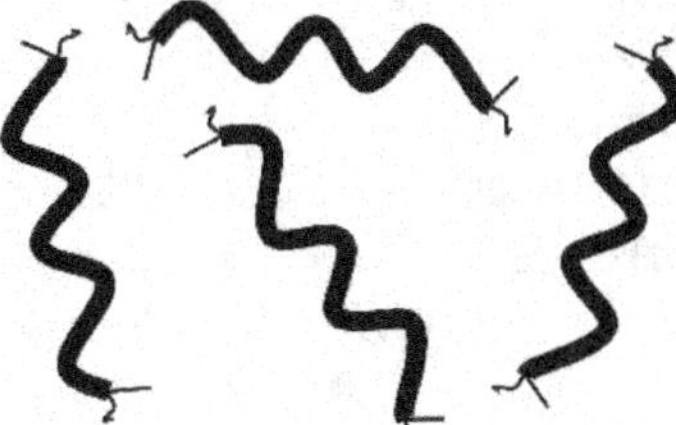

Fig. 16.32: Treponema pallidum

The primary symptoms of syphilis are sores or open wet **ulcer**, which is called a chancre, around the mouth or genitals. These symptoms disappear in about 10 to 90 days.

The secondary symptoms appear after about 6 months of infection. There may be headaches, slight skin rash on hands and feet. These symptoms also disappear without treatment.

The secondary symptoms are followed by a long period of latency or dormancy. There are no symptoms during this period.

The tertiary symptoms of syphilis infection are problems with the heart, brain, and nerves that can result in paralysis, blindness, deafness and even death

Treatment
Syphilis can be treated by antibiotics, which kill bacteria in the body. Doxycycline, erythromycin, tetracycline or penicillin may be used as antibiotics.

Cambridge 5090 syllabus specification 16(dd) discuss the spread of human immunodeficiency virus (HIV) and methods by which it may be controlled.

Human Immuno-deficiency Virus (HIV) causes AIDS. HIV is spread from the body of an infected person to an uninfected person by
- sexual contact with an HIV infected partner
- by sharing injection needles with an HIV infected person
- by blood transfusion from an HIV infected person.

The spread of HIV can be controlled by
- screening of blood to test for HIV before transfusion
- use of disposable needles and syringes
- use of condoms or femidoms during sexual contact
- avoid sexual contact with HIV infected people
- testing of people for HIV and subsequent monitoring and counselling of HIV positive people to prevent them from spreading HIV to non-infected people.

> *Cambridge 5090 syllabus specification 17(a) describe the difference between continuous and discontinuous variation and give examples of each;*

Continuous Variation

Characters showing continuous variation are:

* ❖ controlled by many genes, each gene making a small contribution to the phenotype.
* ❖ there is considerable environmental influence.
* ❖ the characters show a continuous range of variation, with many intermediate phenotypes between the two extreme phenotypes.

Height in humans is an example of a phenotype showing continuous variation. It is controlled by many genes and is influenced by environmental factors like diet, hygiene and exercise. There is a continuum of height between the shortest and the tallest person. If a frequency distribution graph is plotted then it shows a normal distribution (bell shaped pattern), as shown in figure 17.1.

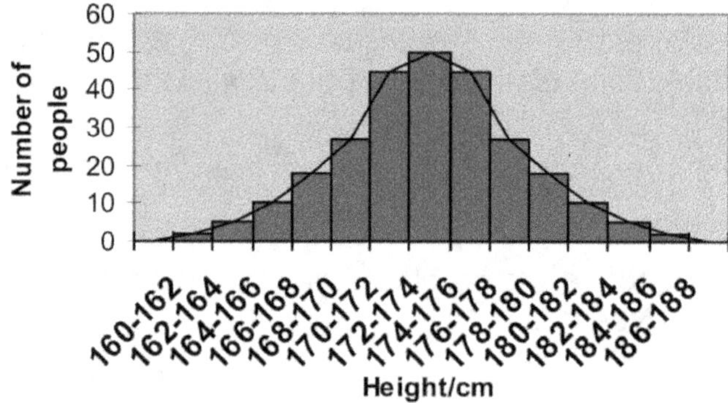

Fig. 17.1: Continuous variation in height of humans

Characters like blood pressure, weight, heartbeat rate, finger length in humans can be measured in some way and are therefore called quantitative characters and shown continuous variation.

Discontinuous Variation

In discontinuous variation, there are clear-cut differences in the forms of various characters. These characters show qualitatively different traits such as white or coloured, wrinkled or smooth seed coat, tall or dwarf plants. Tongue roller or non-rollers, blood groups in humans are examples of discontinuous variation. There are no intermediates between the two forms. The differences between the two phenotypes are determined by alternative forms, or alleles, of a single gene and no measurements are needed to distinguish between them.

Figure 17.2 shows the distribution of height in pea plants. There is a clear-cut difference between the dwarf plants and the tall plants.

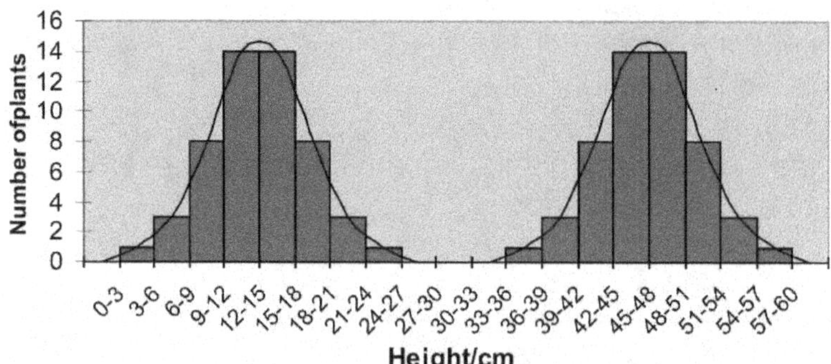

Fig. 17.2: Discontinuous variation in height of pea plants

Chromosomes are thread-like structures found in the nuclei of cells during cell division. The chromosome is a complex of a long and highly folded DNA molecule and proteins. The structure of a chromosome is shown in figure 17.3.

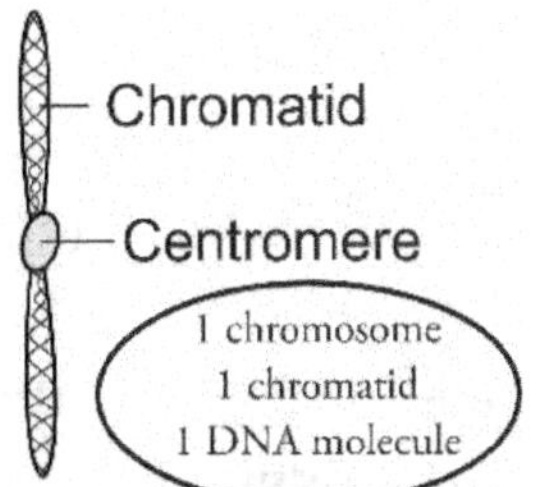

Fig. 17.3: A single chromosome

- DNA is a long molecule found in the nucleus of all cells. In bacteria the DNA lies in the cytoplasm.

- The DNA is divided into sections called genes, as shown in figure 17.4.

- Each Gene is responsible for making a specific protein, as shown in figure 17.4.

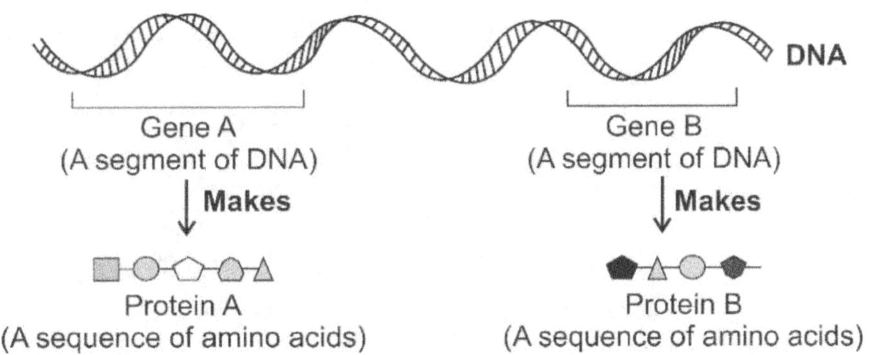

Fig. 17.4: Genes make proteins

Each human chromosome has about a 1000 genes on an average. A copy of each gene is made and distributed to daughter cells during cell division. The gametes carry these genes to the next generation in sexually reproducing organisms. In asexually reproducing organisms the genes are passed on to the next generation by mitosis.

Figure 17.5 illustrates the formation of a chromosome with two copies of each gene. Each gene occupies a specific location on the chromosome, called as the gene locus.

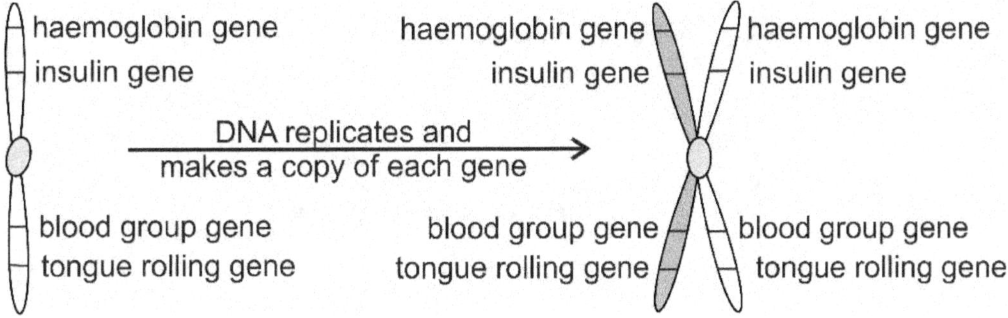

Fig. 17.5: Genes on a chromosome

A gene is a unit of inheritance. A gene can exist in two or more different forms. These alternate forms of a single gene are referred to as alleles.

For example, the haemoglobin gene is used to make haemoglobin. In most people it makes normal haemoglobin. However, some people have a different form of the haemoglobin gene which make a slightly different form of haemoglobin that causes Thalassaemia. Other people may have yet another form of the haemoglobin gene that makes another form of haemoglobin that causes sickle cell anaemia. So, the haemoglobin gene has three different forms. These different forms of the haemoglobin gene are called as alleles.

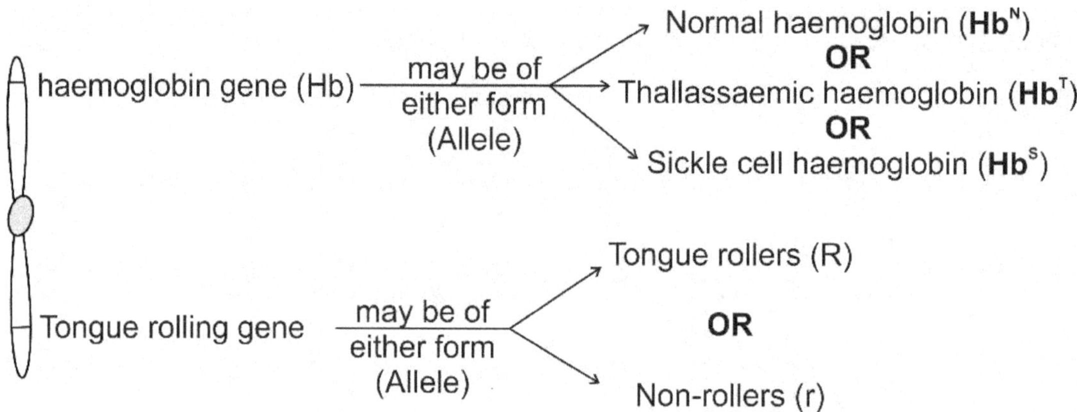

Fig. 17.6: Alleles are alternate forms of a gene

Example of Monohybrid Inheritance

Monohybrid inheritance is the inheritance of a single characteristic. It involves the study of inheritance of a **single gene having two alleles.**

The gene for height can exist in two forms – **T** (allele for tallness) or **t** (allele for dwarfism) thus the alleles of the gene for height are **T** and **t,** as shown in figure 17.7.

Let's consider an example of a Monohybrid cross involving the inheritance of height in pea plants. This character or gene has two alleles Tall (T) and Dwarf (t). It is conventional to represent the dominant allele by a capital letter and the corresponding recessive allele by the simple form of the same letter.

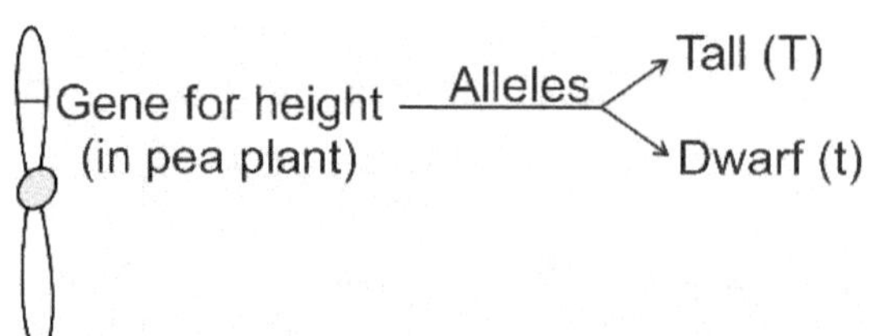

Fig. 17.7: Alleles for height gene in pea plants

Figure 17.8 shows the inheritance of height in pea plants.

- **Genotype:** The combination of alleles found in an organism. For example TT, Tt or tt are the possible genotypes of the gene for height.
- **Phenotype:** The observable physical characteristics of an organism. For example Tall or Dwarf plants.
- **Homozygote:** A genotype consisting of similar alleles. For example TT and tt are both homozygous.
- **Heterozygote:** A genotype consisting of two different alleles. Tt is heterozygous.
- **Dominant allele:** The allele which expresses itself in the phenotype, in both, homozygous and heterozygous genotypes. For example the allele for tallness expresses itself in TT and Tt genotypes.
- **Recessive allele:** The allele expresses itself in the phenotype, only in homozygous conditions but, not in heterozygous condition. tt is a dwarf plant. In heterozygous condition, Tt, the recessive allele(t) is present, but is not expressed in the phenotype.

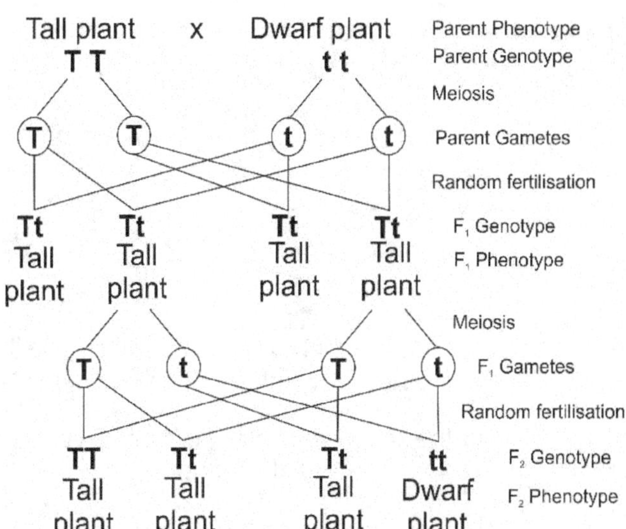

Fig. 17.8: Monohybrid cross in pea plant

Cambridge 5090 syllabus specification 17(i) *predict the results of simple crosses with expected ratios of 3:1 and 1:1, using the terms homozygous, heterozygous, F_1 generation and F_2 generation;*
Cambridge 5090 syllabus specification 17(j) *explain why observed ratios often differ from expected ratios, especially when there are small numbers of progeny;*

Predicting the results of genetic crosses

An analogy of a coin can be used to predict the results of simple crosses, as shown in figure 17.9. The distribution of alleles to the offspring is random, just like the flipping of a coin.

Increasing the number of flips will get us closer to the expected 3:1 F_2 ratio, as the expected ratio is due to chance or randomness.

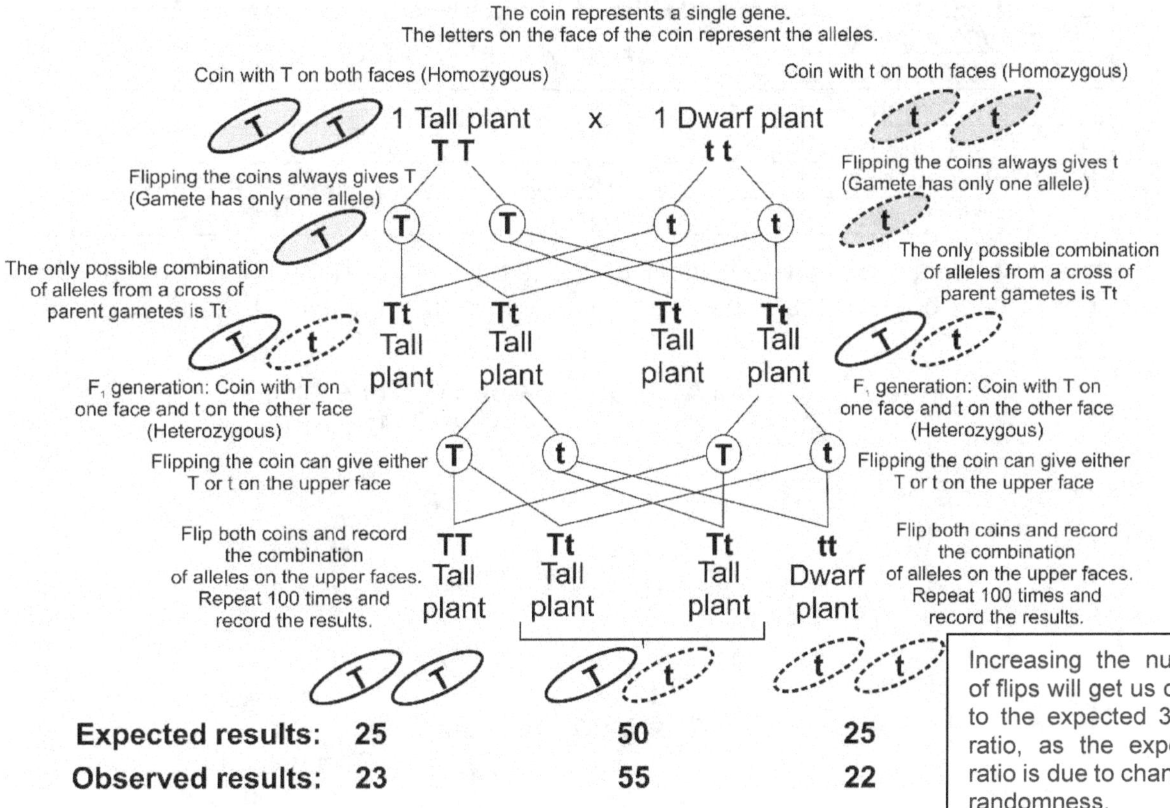

Fig. 17.9: Predicting the results of genetic crosses

Test cross or back cross

A test cross or back cross is done to determine the genotype of an individual showing the dominant phenotype. For example the tall plant may be homozygous (TT) or heterozygous (Tt). Its genotype can be determined by crossing it with a homozygous recessive (tt) dwarf plant, as shown in figure 17.10.

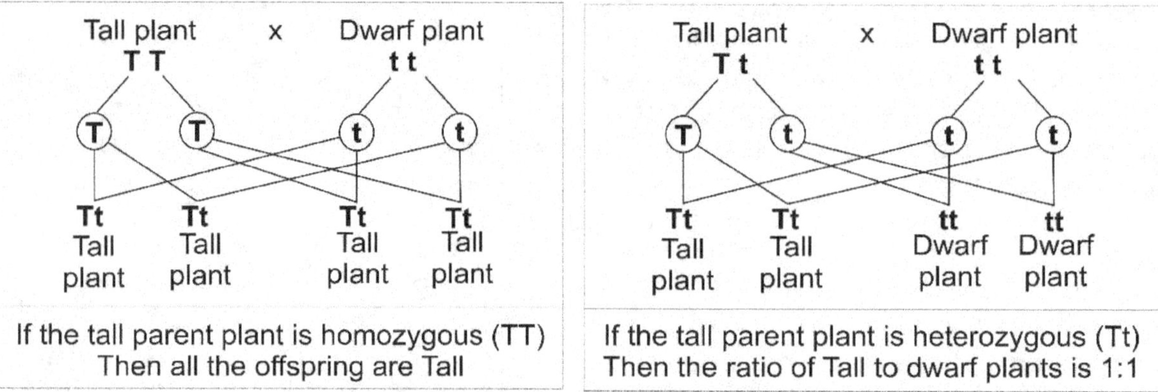

Fig. 17.10: Test or back cross

Cambridge 5090 syllabus specification 17(k) explain codominance by reference to the inheritance of the ABO blood group phenotypes (A, B, AB, O, gene alleles I^A, I^B and I^O);

The ABO blood group is controlled by a single gene with three alleles, I^A, I^B and I^O.

Alleles	Type of allele
I^A	Dominant (Expresses itself in the phenotype in homozygous and heterozygous condition)
I^B	Dominant (Expresses itself in the phenotype in homozygous and heterozygous condition)
I^O	Recessive (Expresses itself in the phenotype only when in homozygous condition)

The possible genotypes and phenotypes of the ABO blood group are listed in the following table.

Genotype	Phenotype (Blood group)	Gene interaction
$I^A I^A$ (Homozygous)	*Group A*	Complete Dominance
$I^A I^O$ (Heterozygous)		
$I^B I^B$ (Homozygous)	*Group B*	Complete Dominance
$I^A I^O$ (Heterozygous)		
$I^A I^B$ (Heterozygous)	*Group AB*	**Codominance**
$I^O I^O$ (Homozygous)	*Group O*	Recessive

In complete dominance, one allele expresses itself while the other allele does not when found in heterozygous condition. In codominance, both alleles express themselves equally in heterozygous condition, as shown in figure 17.11.

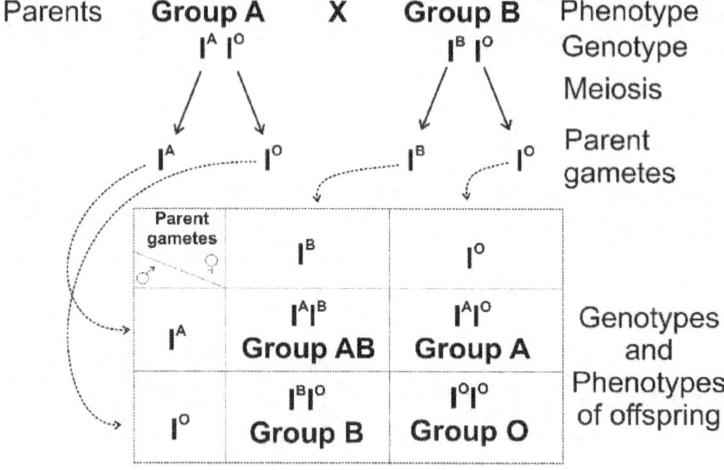

Fig. 17.11: Inheritance of blood group alleles

Mutations are changes in the structure of the gene or in the number of chromosomes. The rate of mutations are increased by mutagenic agents such as X-rays, gamma rays, ultraviolet radiation and chemicals in cigarette smoke and mustard gas.

A **gene mutation** in the haemoglobin gene will result in the structure of the haemoglobin protein to change. This is responsible for causing sickle cell anaemia. The haemoglobin will crystallize and the red blood cells become sickle shaped (crescent), as shown in figure 17.12. Many red blood cells are destroyed by the spleen and the person becomes anaemic (lacks haemoglobin).

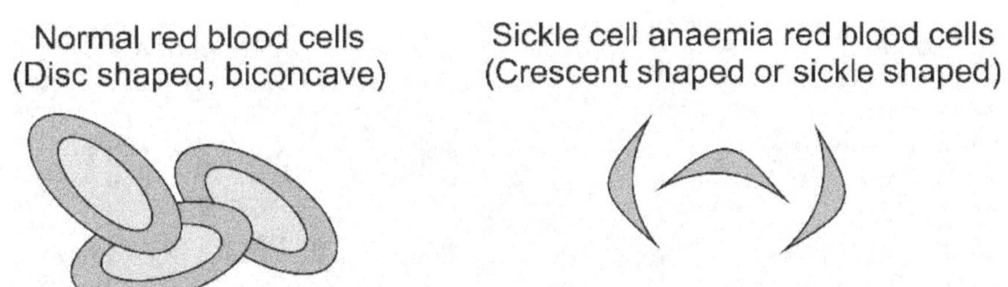

Fig. 17.12: Sickle cell anaemia

A **chromosomal mutation** results in a change in the number of chromosomes in the cell. Down's syndrome is caused when a gamete, formed during meiosis, has an extra chromosome. The fusion of this gamete with a normal gamete results in the zygote having one extra chromosome, as shown in figure 17.13.

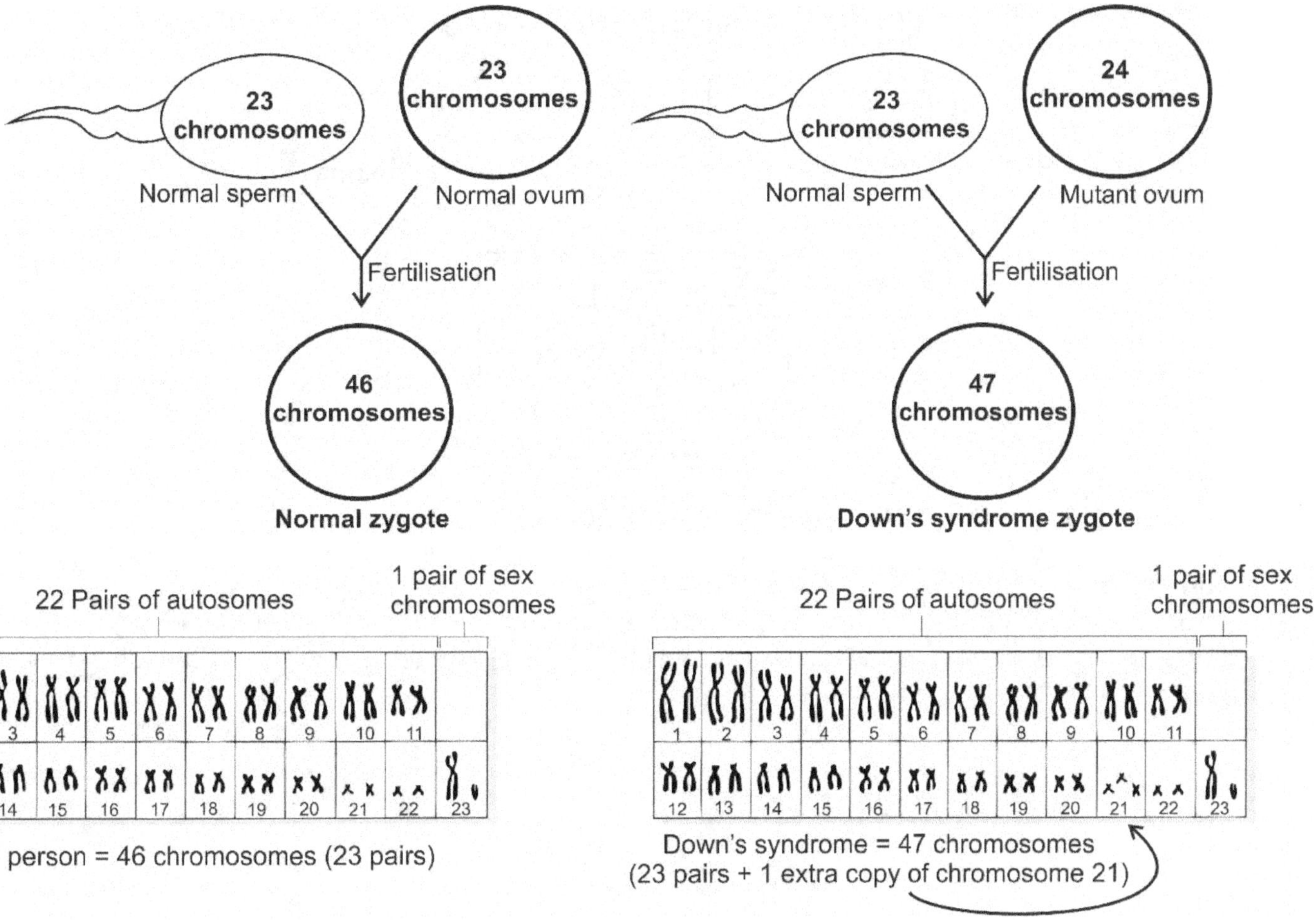

Fig. 17.13: Down's syndrome

Sex determination in Humans

- There are 23 pairs of chromosomes in humans.
- 22 pairs are not involved in the determination of sex in humans. These chromosomes are called **autosomes**.
- The 23rd pair of chromosomes is involved in sex determination and is called **sex chromosomes or allosomes**.
- In humans, there are two types of sex chromosomes – the X chromosome and the Y chromosome.
- The karyotype XY gives rise to males and the karyotype XX gives rise to females, as shown in figure 17.14.

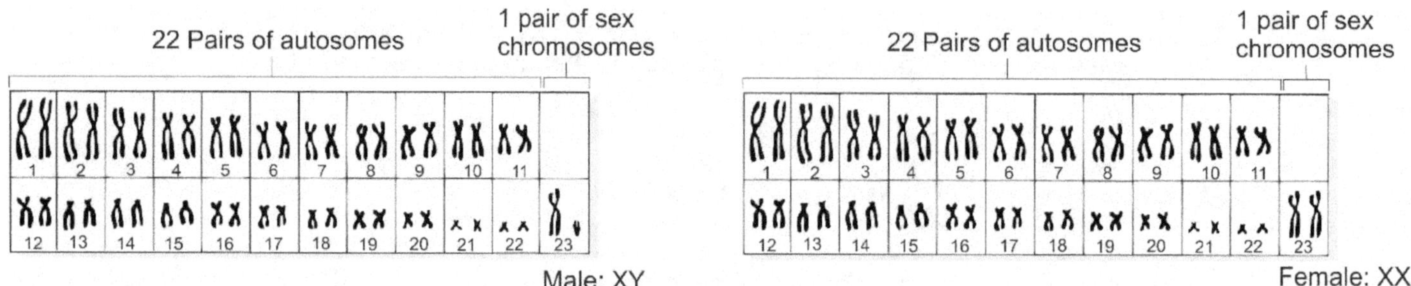

Fig. 17.14: Karyotypes of male and female

Figure 17.15 shows the segregation of sex chromosomes in human gametogenesis.

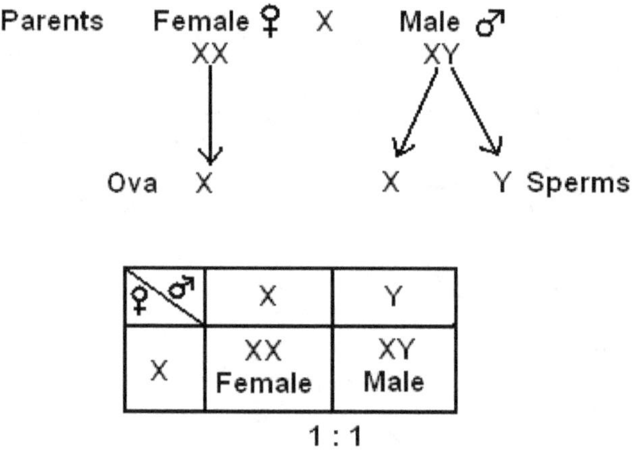

♀ \ ♂	X	Y
X	XX Female	XY Male

1 : 1

Fig. 17.15: Sex ratio in humans

- Female produces ova containing one X chromosome.
- Male produces 50% sperms containing one X chromosome and 50% of sperms containing one Y chromosome.
- Thus, the expected ratio of offspring should be 50% male (XY) and 50 % female (XX).
- Thus, the male gamete determines the sex of the baby. The female is homogametic – producing one type of gamete.

There is variation among the offspring of any species. Some organisms have adaptations that help them to use the resources of the environment better than others. For example, consider a cactus in the desert. Due to variation, some plants have very short roots, which are not widespread. Other plants of the same species have plants with long, widespread roots. The plants with longer roots have a better chance of surviving to a reproductive age. They pass on the genes for long roots to their offspring. So, **over many generations** the population will have cacti with long roots. This is called natural selection. It plays a role in making species better adapted to the environment. It also helps in forming new species.

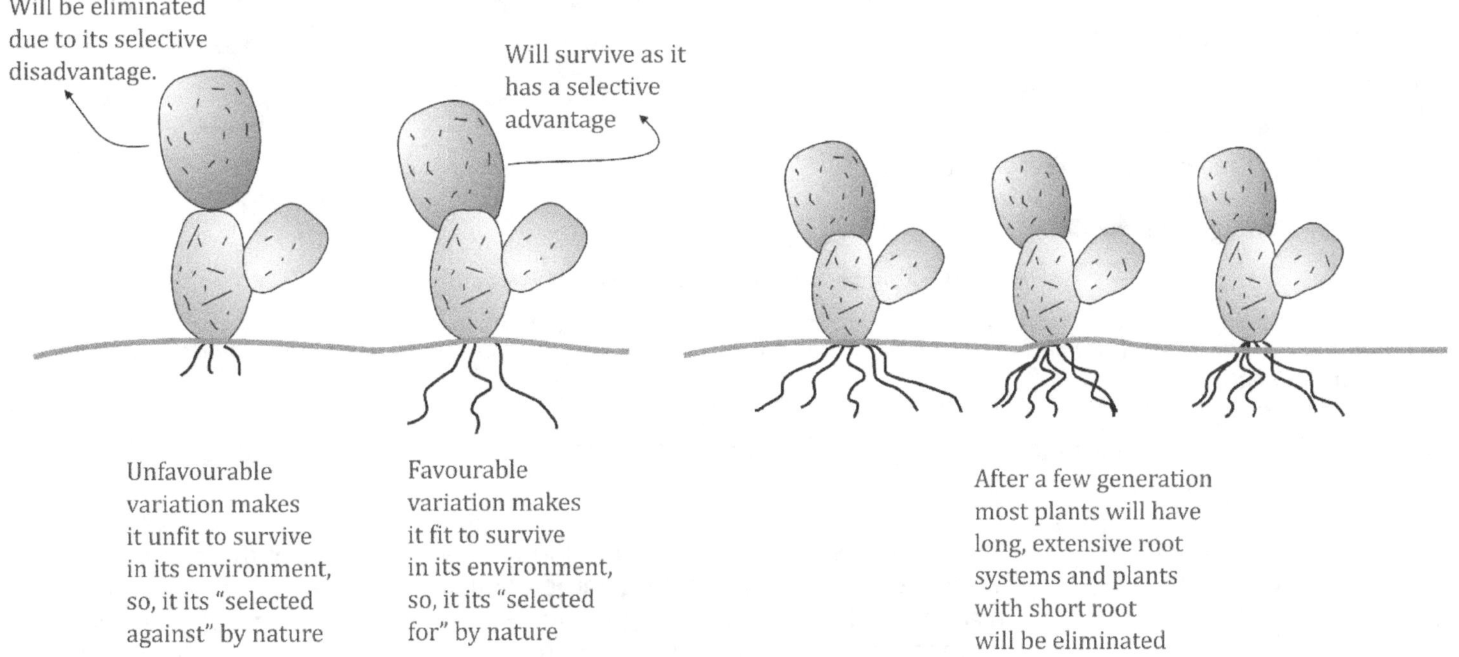

Fig. 17.16: Natural selection leading to adaptation

Natural selection leads to formation of new species (evolution)

Consider the sequence of events that lead to formation of new species

1) There is a population of moths living in all parts of a tropical forest. The conditions in the forest have been very similar for millions of years and the moths are well adapted to these conditions. The population has a single gene pool and the individuals are able to interbreed freely- we say that there is a regular 'gene flow'.

2) Climatic changes occur, much of the forest dies out and are replaced by scrub or desert, where moths cannot survive. Some of the forest survives as two separate areas- forest A and forest B on the mountainsides. Each forest will now have its own population of moths. The desert isolates the populations in both forests (Geographical isolation), as shown in figure 17.17. Gene flow between the two populations is now prevented.

3) Since environmental conditions in both forests are different, each population of moths will be subjected to different selective pressures. Natural selection would favour genetically new kinds of moths, which would be better adapted to the new environmental conditions. This may cause the two populations of moths to become genetically different enough to be reproductively isolated, so producing a new species.

4) Climate change occurs again and the desert again becomes a forest. The two populations of moths are now able to meet up (mix). But the gene pool of each population has diverged to such an extent that they have become reproductively isolated and are no longer capable of interbreeding. Two different species have now evolved.

Fig. 17.17: Formation of new species by natural selection

Cambridge 5090 syllabus specification 17(o) describe the role of artificial selection in the production of economically important plants and animals;

Artificial selection can be used to produce a large number of plants or animals with the desirable characteristics. The process is outlined in figure 17.18.

Identify useful characteristics in an animal or plant. e.g. cows with high yield of milk, or plants with high yield of food $\longrightarrow$ Select organisms with the useful characteristics for breeding. $\longrightarrow$ Select offspring with the useful characteristics to produce the next generation. $\longrightarrow$ Repeat the process for many generations to produce organisms with the desirable characteristics.

Fig. 17.18: Artificial selection

Cambridge 5090 syllabus specification 17(p) explain that DNA controls the production of proteins;
Cambridge 5090 syllabus specification 17(q) state that each gene controls the production of one protein;
Cambridge 5090 syllabus specification 17(r) explain that genes may be transferred between cells (reference should be made to transfer between organisms of the same or different species);
Cambridge 5090 syllabus specification 17(s) explain that the gene that controls the production of human insulin can be inserted into bacterial DNA;
Cambridge 5090 syllabus specification 17(t) understand that such genetically engineered bacteria can be used to produce human insulin on a commercial scale;

Genetic engineering

- Genetic engineering is the process by which genes from one organism are transferred into the cells of another organism.
- The gene may be transferred into another organism of the same species or into the cells from other species.
- The gene controls the production of a specific protein in the cell.
- The gene for human insulin can be transferred into bacterial cells. The bacterial cell then produces insulin, as shown in figure 17.19.

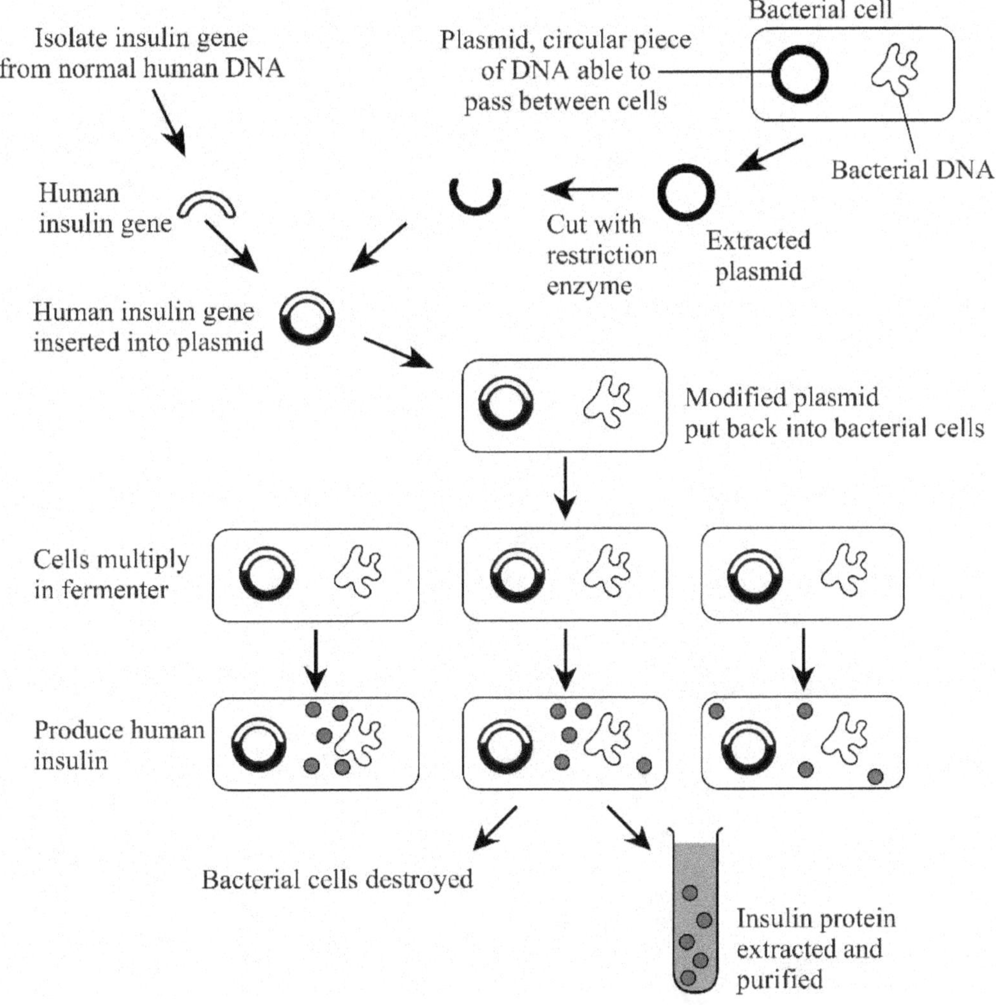

Fig. 17.19: Production of insulin by genetically engineered bacteria

Cambridge 5090 syllabus specification 17(u) discuss potential advantages and dangers of genetic engineering.

Potential benefits of genetic engineering	Potential risks of genetic engineering
Genetically engineered plants may be disease and drought resistant. They may also produce more food for the rapidly growing world population.	May have unpredictable health risks. We cannot know the consequences of transferring genes across the species barrier. Proteins produced may be dangerous (toxic or initiate allergic responses), either directly, or indirectly through metabolic processes.
Genetically engineered plants are more cost-effective as less money has to be spent on fertilisers and insecticides.	Could increase herbicide and pesticide use. The genes used for increased herbicide resistance in GM crops may spread to other plants, including weeds. This may allow for the development of resistant 'super weeds'.
Genetically engineered plants could benefit human health. Genetically engineered Golden Rice – has a high vitamin A content, and reduces the number of people suffering Night blindness.	Could reduce biodiversity. It has been argued that the use of GM crops reduces the chance of other plants or insects existing within the crop, and hence reduces biodiversity.